American Institute of Chemical Engineers

Topical Conference on Process Intensification and Clean Technology 2005

AA003261

Held at the 2005 AIChE Spring National Meeting

April 10-14, 2005
Atlanta, Georgia, USA

Printed from e-media with permission by:

Curran Associates, Inc.
57 Morehouse Lane
Red Hook, NY 12571
www.proceedings.com

ISBN: 978-1-60423-471-8

American Institute of Chemical Engineers

Topical Conference on Process Intensification and Clean Technology
2005

TABLE OF CONTENTS

Session 3 - Novel Separation Technology

Chair: Roshan Jachuck
Vice Chair: Jo Rogers

Session 4 - Microreaction Technology

Chair: Roshan Jachuck

1a Keynote Presentation

John Burns

Opportunities and Challenges within Intensified Reactor Techniques

1b The Crucible for Differential Scanning Calorimetry as a Microreactor for the Investigation of Oxidation reactions and Hydrogenations

Pierre Reuse

Differential Scanning Calorimetry (DSC) is a well-known technique of analysis since almost 30 years. Its application in the domain of the chemical process safety has allowed a routine determination of the thermal potential of reactants, reaction masses and products leading to a significant increase of the safety level in the industrial world.

Most chemicals produce gaseous compounds during their decomposition. It has been necessary to develop, for technical and for occupational safety reasons, a pressure tight crucible. The inner volume of these crucibles is around 20 microliters and the typical amount of sample used is 5 mg. These crucibles are able to with-stand 200 bars at 400 C.

We developped a new device allowing us to work with a controlled atmosphere in these crucibles. Indeed it is now possible to close the crucibles under argon, oxygen or hydrogen and within a pressure range of 1 to 20 bars.

The results that will be presented will show that this new preparation method gives a powerful insight in oxidation reactions and hydrogenations at the microgram scale.

For example by varying the initial pressure of oxygen it is possible to stop the oxidation reaction at various thermal conversion.

The high surface to volume ratio of these samples makes possible the micro-hydrogenation as batch reaction in a DSC-crucible. This tool could be used as high-throughput catalyst screening device. The results (obtained with this method) of the determination of the energy of the hydrogenation of a triple bond are completely comparable tothe results obtained in a reaction calorimeter.

1c Hydrodynamic Cavitation: A Process Intensification Technology For The Future

Aniruddha B Pandit, Girish V Ambulgekar, Parag R Gogate

Cavitation can be effectively used for variety of industrial applications such as chemical synthesis (in both homogenous and heterogeneous systems by way of increase in the rate and selectivity of many chemical reactions); wastewater treatment (degradation of many of the biorefractory/complex chemicals); textile processing (for improving the efficacy of dyeing technique and for generation of clean effluents); biotechnology (cell disruption for release of intracellular enzymes and foam control in bioreactors); crystallization etc. These applications, in particular, the application of cavitation for intensification of chemical reactions, are attributed to the generation of local conditions of high pressures and temperatures (hot spots), turbulence with associated liquid circulation and generation of free radicals. In the past, cavitation generated by the use of ultrasound (acoustic cavitation) has been studied extensively for the development of industrial scale applications but it has number of technical and economical processing limitations. Hydrodynamic cavitation, an alternative technique for the generation of cavitating conditions and studied in details only recently, appears to be much easier for the scale up and industrial scale applications. In the present work, the basic aspects of hydrodynamic cavitation, optimization of the operating parameters and few experimental illustrations, which clearly indicate the superiority of hydrodynamic cavitation as compared to acoustic cavitation and/or conventional reaction techniques, have been discussed.

Hydrodynamic cavitation can simply be generated by the passage of the liquid through a constriction such as throttling valve, orifice plate etc. and the geometry of the constriction strongly influences the intensity of the cavitating conditions as well as the zone of influence of cavitation. A careful design of the system allows generating conditions similar to acoustic cavitation thereby enabling different applications, which have been successfully carried out using acoustic cavitation phenomena but at much lower energy inputs as compared to sonochemical reactors.

The discussion related to theoretical modeling of the bubble dynamics equations briefly explains the approach for the quantification of the collapse pressure generated at the end of cavitation events using either a single cavity or a more realistic cavity cluster considerations. Some recommendations have been made for the selection of the optimum operating conditions based on the predictions of the simulations and subsequent comparison with experimental results reported in the literature.

The experimental work concentrates on the chemical synthesis applications and synthesis of bio-diesel has been considered as the model reaction. Today "bio-diesel" is the term applied to the esters of simple alkyl fatty acids used as an alternative to petroleum based diesel fuels. The hydrodynamic cavitation reactor used in the work consists of a reservoir or a collecting tank with 10 L capacity that is connected to the pump (capacity of 1.5kW) which allows re-circulation of the contents through a mainline housing an orifice plate (different configurations of the holes or the orifice plates can be inserted in the main line) and a bypass line (for controlling the inlet pressure and the flow rate into the cavitation chamber). A cooling water jacket is also provided to the reactor for controlling the temperature of the system as usually temperature rises due to the mechanical heat dissipation into the system and higher operating temperature are detrimental for the cavitational reactions. The performance of the hydrodynamic cavitation reactor has been quantified in terms of cavitational yield, which gives the amount of the product generated per unit electrical energy supplied to the reactor. For comparison purpose, the reaction has been also carried out using an ultrasonic bath (capacity of 55mL, frequency of irradiation as 20kHz and power dissipation of 120W) and using the conventional approach of stirring and reflux. Cavitational yield was calculated for each of these operations.

It has been observed that for the model reaction, the cavitational yield for hydrodynamic cavitation reactor is order of magnitude higher as compared to the sonochemical reactors and the conventional approach. Specifically, that hydrodynamic cavitation is about 40 times more efficient as compared to the acoustic cavitation and 160 to 400 times more efficient as compared to the conventional agitation/heating/refluxing method. To, get some more insight into the comparison of the efficacy of the two cavitation techniques, cavitational yield have been also estimated for oxidation of toluene, o-xylene, m-xylene, p-xylene, mesitylene, o-nitrotoluene, p-nitrotoluene, m-nitrotoluene, o-chlorotoluene and p-chlorotoulene and it has been observed that hydrodynamic cavitation indeed is much more efficient as compared to acoustic cavitation.

The present work has enabled us to conclusively establish the efficacy of the hydrodynamic cavitation reactors as compared to the acoustic counterparts at least for the applications explored in this work. It can be said that the hydrodynamic cavitation reactors offer immediate and realistic potential for industrial scale applications as compared to the sonochemical reactors and the scale up of these reactors is comparatively easier as vast amount of information about the fluid dynamics downstream of the constriction is readily available and the operating efficiency of the circulating pumps which is the only energy dissipating device in the system is always higher at large scales of operation.

HEAT TRANSFER ENHANCEMENT AND FOULING RATE REDUCTION IN CHEMICAL INDUSTRY PROCESS HEATERS THROUGH DIMPLING OF THE PRODUCT TUBES

Yaroslav Chudnovsky, Ph.D., Aleksandr Kozlov, Sc.D.
Gas Technology Institute, Des Plaines, IL

Abstract

Many existing heat transfer enhancement techniques are effective to increase heat transfer rates, however, in most cases this increase is invariably accompanied by an associated increase in pressure drop. For heaters firing "dirty" fuels the increased potential for fouling of the tubes must also be considered.

GTI has identified a technology that increases heat transfer without any significant increase in pressure drop. It involves the use of special 3-D profile (dimples) on the heat transfer surface. The enhancement occurs due to the vortex flow pattern that is generated in each cavity and provides intensive heat and mass transfer between the surface and heating media. Moreover, it is expected that the surface dimpling will not increase (and may even reduce in some cases) the fouling rate as a result of the columnar vortices that are formed in the dimples and extensively evacuate deposits from the dimpled surface.

This paper will present the results of initial laboratory experiments at GTI comparing Dimpled Tubes with the type of finned tubes used in the Chemical Industry. The initial results show the following summarized benefits:

- Relative heat transfer coefficient ($Nu_{dimpled}/Nu_{finned}$) ~ 3.0
- Relative pressure drop ($Eu_{dimpled}/Eu_{finned}$) ~ 0.4

A field trial at Industrial Partner's refinery is underway. Preliminary results are expected in the second quarter of 2005.

Introduction

The overall objective of this work is to develop a competitive and cost-effective technology for significantly improving the energy efficiency of fired process heaters in the Chemical Industry. Specific target includes a 3-5% increase in overall heater thermal efficiency with no significant increase in pressure drop and a reduction in fouling rate. A number of approaches can be used to increase heat transfer rates in the convective section of fired process heaters. Majority of these approaches are effective in increasing the heat transfer rates, however, this increase is invariably accompanied by an associated increase in convective section pressure drop as well as, for heaters firing 'dirty' fuels, increased fouling of the tubes - both of which are highly undesirable.

GTI has identified an approach, which will increase heat transfer rates without a significant increase in pressure drop and a reduction in fouling rate. It involves the use of special 3D surface profile on the heated tubes in the convection section (generally, spherical dimples pattern with certain arrangement). This technology is known as Vortex Heat Transfer

Enhancement (VHTE). Compared to other types of heat transfer enhancement approaches, as illustrated in Figure 1, the proposed approach achieves very high heat transfer rates at the low pressure drops. The red line in Figure 2 represents the equality of heat transfer gain and pressure drop penalty associated with this gain reached by majority of the heat transfer enhancement techniques. Figure 2 shows that VHTE provides a better balance of pressure drop and heat transfer than most other available techniques.

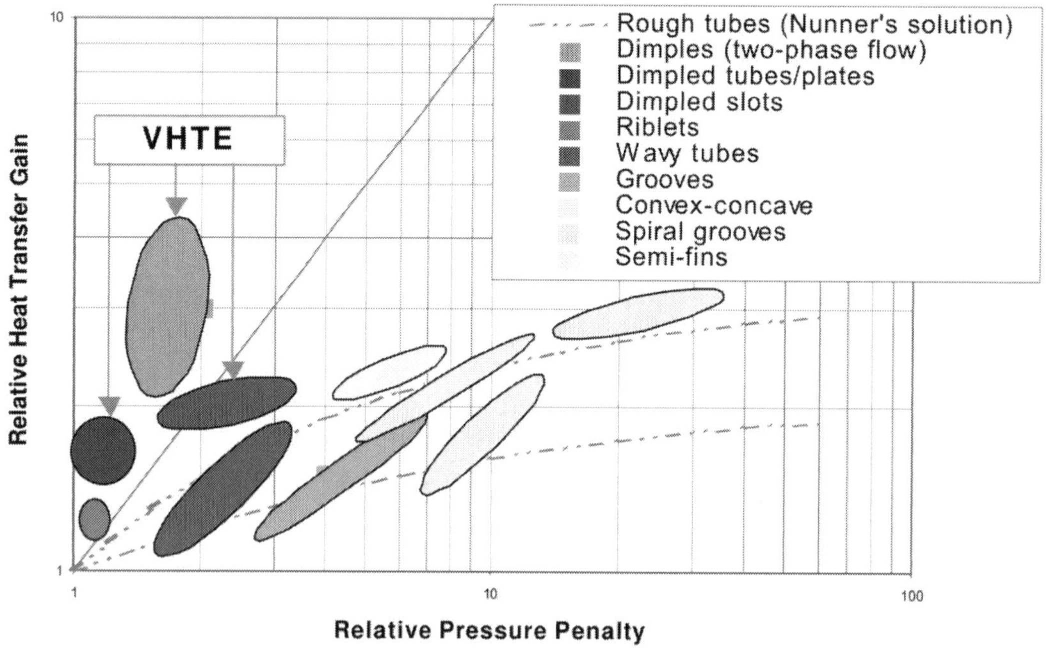

Figure 1. Efficiency of various heat transfer enhancement methods

Studies [1] have confirmed that while flowing a gaseous medium over surfaces with dimpled surface profile, each dimple works as a "vortex generator" that intensifies the rate of convective heat and mass transfer between the dimpled surface and the core of the gaseous medium (see Figure 2). The following benefits over bare surface were established for dimpled surfaces arranged in a tube bundle [2]:

- Increased heat transfer coefficient up to 20-30%
- No significant increase (or even reduction in some cases) in pressure drop

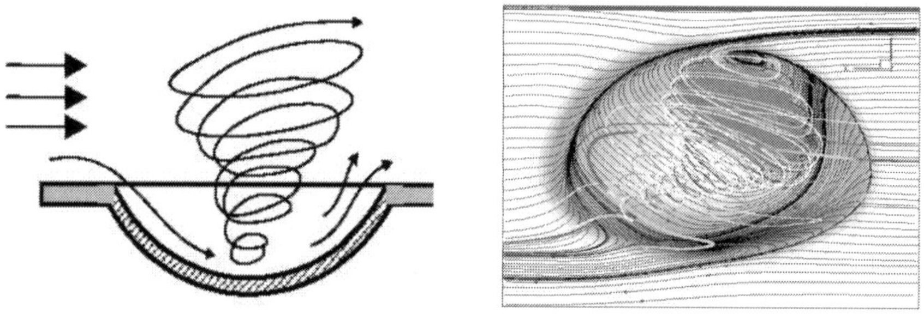

Figure 2. Mechanism of swirl formation in a spherical cavity [3] and example of CFD flow pattern [4]

Application Concept

Implementation of the VHTE technology into industrial applications will require the replacement of all the tubes in the existing (or new constructed) convective section with VHTE enhanced tubes (including shock tubes). The general layout of the originally proposed application concept is shown in Figure 3 below.

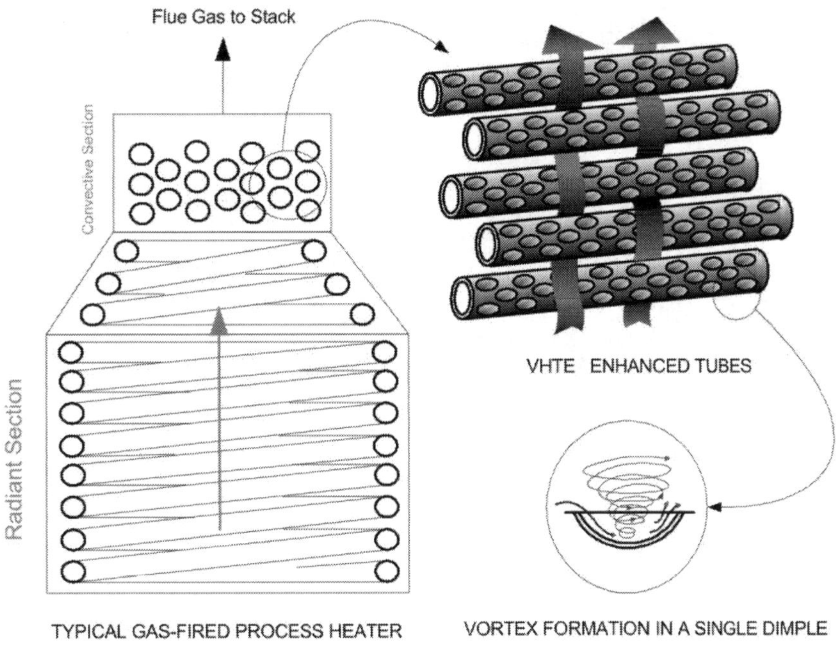

Figure 3. Application concept of VHTE enhancement of a fired heater convection section

Development Status

The key hurdles of the development are:

- The risk of applying a new technology into an operating fired process heater. This risk was minimized by involving industrial partners with a strong presence in the chemical industry and which complement the effort by proving the technical and economic benefits of the technology in the field as well as assisting in identification of suitable application process.
- The risk that laboratory results will not adequately represent full-scale performance. This risk was reduced by thorough laboratory testing and the use of CFD modeling.
- Convincing potential users that the benefits of dimpled-tube technology outweigh its costs. This will be addressed by examples that illustrate satisfactory monetary rates of return upon successful field trial.

During the continuing development three major steps were undertaken so far: CFD modeling to develop the efficient dimpled surface profile including experimental validation of the developed CFD model [5]; bench-scale evaluation following the comparative analysis of six test sections consisting of the different types of tubes: bare, finned and dimpled [6]; and field trial system development including design, fabrication and installation at the Industrial Partner's refinery.

Bench-Scale Evaluation

To evaluate bare, finned and dimpled tube banks in terms of heat transfer and pressure drop the Bench-Scale Unit (BSU) along with test sections (replaceable tube banks) were designed and fabricated by a local vendor. Figure 4 represents the general layout of the experimental rig installed at the GTI Applied Research Laboratory in Des Plaines, IL.

Figure 4. General layout of the experimental rig at the GTI laboratory

A test section consists of a rectangular air channel with 20 tubes installed in a staggered arrangement. The pre-heated air (up to 1200°F) flowed over the tubes bank while city water flowed inside the tubes and provided heat removal from the BSU. Six test sections were developed and fabricated for BSU evaluation (see Figure 5 below representing three types of test sections – one bare, one finned and four dimpled sections).

a b c

Figure 5. BSU test sections: **a** – bare (1), **b** – finned (1), **c** – dimpled (4)

Each test section was equipped with pressure measurement taps, thermocouple couplings, and water connection nipples. To minimize heat loss to the ambient environment, the test section was heavily insulated during the testing (see Figure 6).

Figure 6. Test Section Installed in the BSU at GTI

During the test run the following parameters were measured and electronically recorded (for each test section) for further processing and analysis:

- air flow rate, SCFH
- water flow rate, GPM
- air temperature before test section, ^{o}F
- air temperature after test section, ^{o}F
- water inlet temperature, ^{o}F
- water outlet temperature, ^{o}F
- pressure drop across the tube bank, in.WC

The overall heat transfer coefficient and pressure drop across the tube bank for each of the six test sections were obtained. The results for dimpled tubes were compared with bare and finned tubes. For more details see [5].

Heat Transfer and Pressure Drop

Based on temperature and flow measurements the following results were obtained for six test sections. Criterion **Nu** per bare tube diameter (**$Nu=hD/k$**) versus criterion **Re** constructed per inlet flow velocity (**$Re_d=uD/v$**) is given in Figure 7a. Here, **D** is the tube outside diameter, **u** is inlet velocity, **k** is heat conductivity, and **?** is viscosity. Area **A_0** in the figure relates to the outside surface area of the bare tubes and area **A** relates to the outside surface area of the finned tubes. It is obvious that in the **Re** number range of 5000-6500 the heat transfer coefficient is more than twice higher for dimpled tubes as opposed to that for finned tubes. In Figure 7b the criterion **Eu** is presented versus criterion **Re_{max}** that represents the maximal air velocity through the test section. In **Re_{max}** number range of 15,000-25,000 the pressure drop can be reduced up to 3 times by using the dimpled surface.

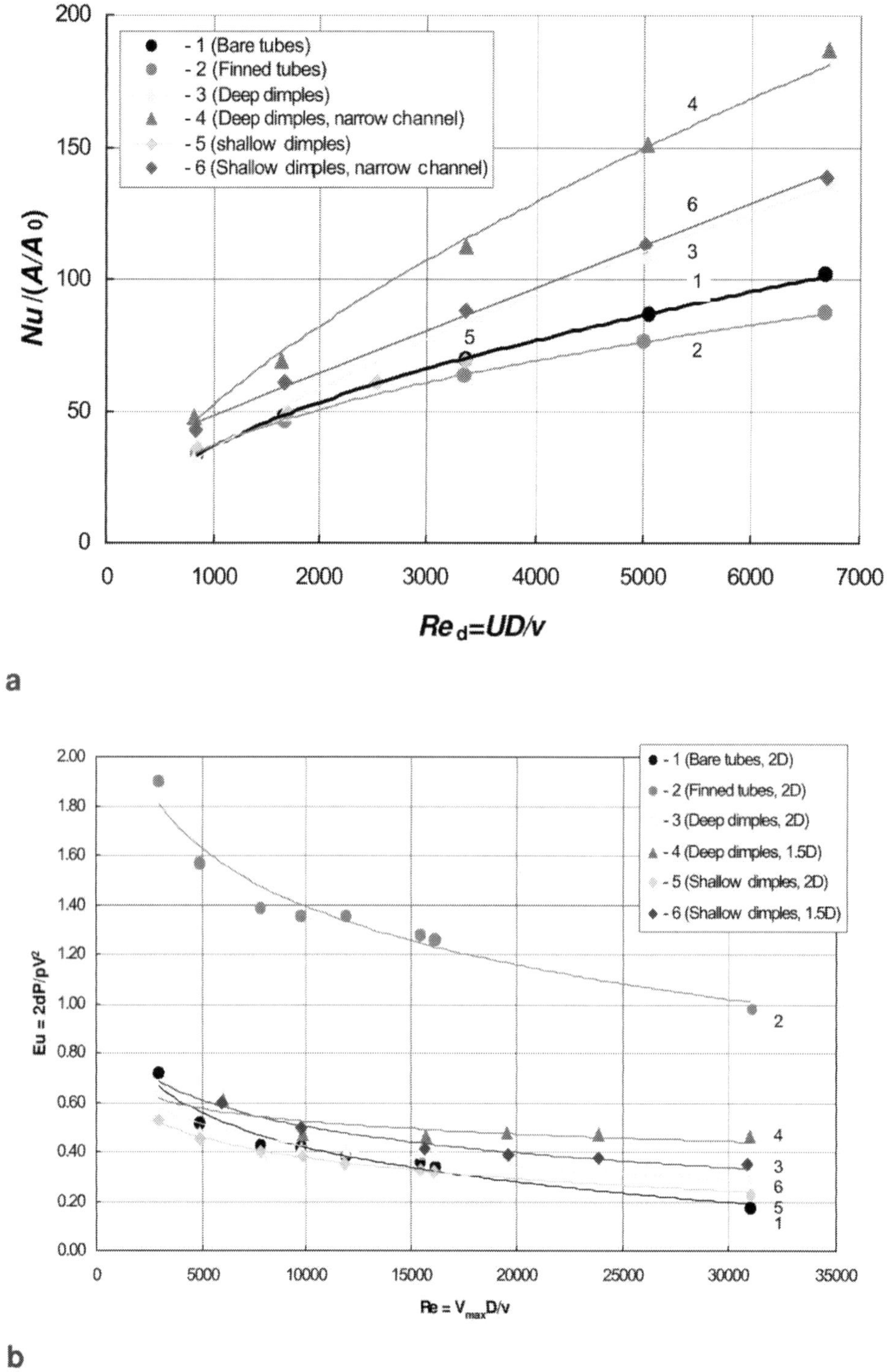

Figure 7. Heat transfer (a) and pressure drop (b) versus air velocity for bench-scale tests

Figure 7 demonstrates the obvious advantage of the dimpled tubes (for all dimple depths and profile densities) over the finned surface. The more details on data processing and error analysis are given in [7].

Field Trial Approach

A slip-stream approach was selected for the upcoming field trial as the most viable way to minimize intrusion into operating unit and risk of loss product. Two ductworks with finned tube and dimpled tube test sections are located in a partial bypass the existing economizer as shown in Figure 8. The ducts were connected to the existing unit through the shut-off guillotines to eliminate, in case of emergency, any negative affect on actual production. The test sections will be put at the similar flue gas and product flow conditions for heat transfer and pressure drop data collection followed by long-term monitoring of the fouling performance.

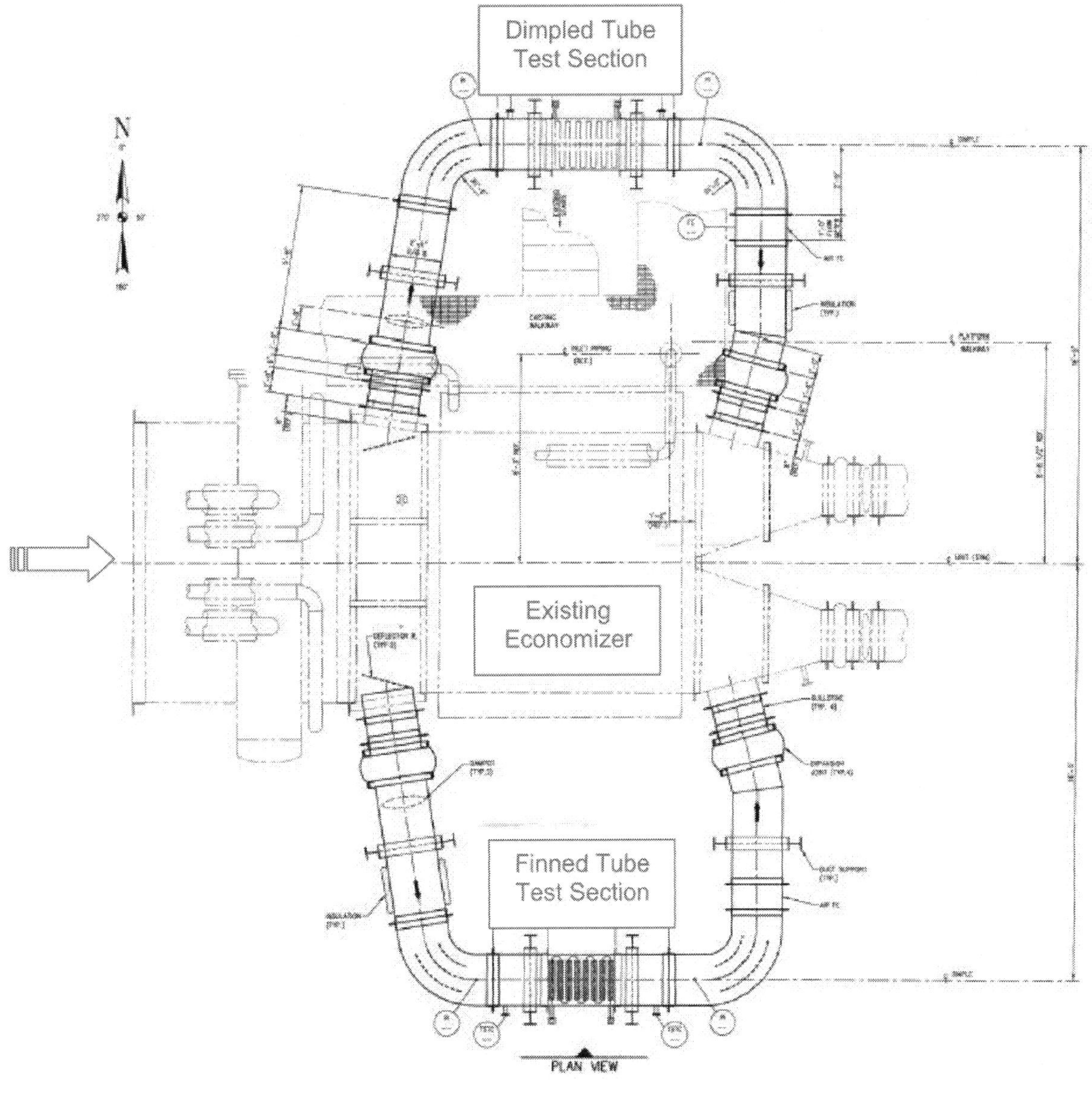

Figure 8. Slip-stream layout of the GTI field trial

Dimpled Tube Development

To meet ASME code the standard tube SA178A was selected for the field trial unit. The dimpled profile developed in the course of bench-scale evaluation was formed on the test tubes surface by an automatically operated tool that was specially designed for this purpose.

The strength of randomly selected tubes was evaluated by means of stressed-deformed state analysis and destructive testing. The destructive testing results demonstrated a secure safety factor. The burst pressure significantly exceeded the operating pressure of the existing economizer. For the follow-on stressed-deformed state analysis, the dimpling geometry was copied directly from the fabricated samples shown in Figure 9.

a 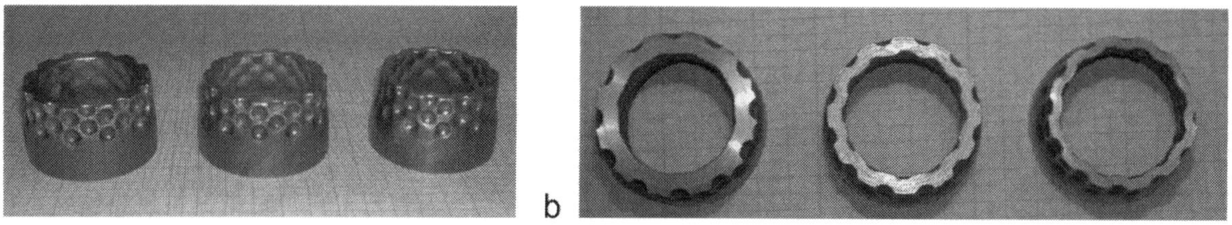 b

Figure 9. Dimpled-Tube Samples Used as a Basis for Deformed-State Analysis:
a – Cuts for Profile Measurement, b – Profiles (Wall Thickness of 0.2", 0.15" and 0.1")

Two values of internal pressure (7,500 psi and 1,400 psi) were considered for all calculations. The first pressure value of 7,500 psi was selected from the destructive test results. The second pressure value, 1,400 psi, is twice the 700-psi working pressure of the tubes. Figure 10 shows part of a finite element stress-deformed state analysis at 7,500 psi.

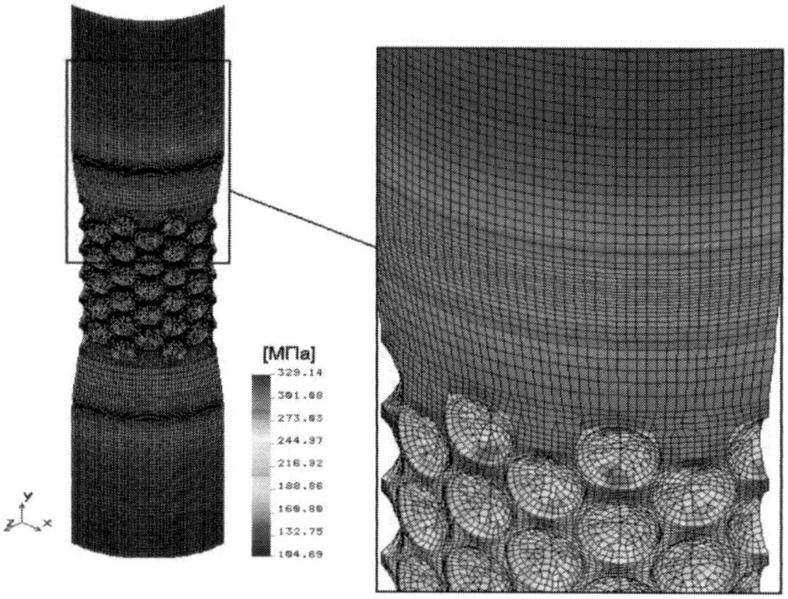

Figure 10. Finite Element Stressed-Deformed State Analysis of a Dimpled-Tube Sample

The yield strength of SA-178A is in the range of 180-255 MPa, so Figure 10 clearly indicates that the obtained maximum stresses exceed the yield point, which means that the material is in its plastic range. Based on the strength analysis performed for 1,400 psi, it is expected that all the stresses will be related to the elasticity range, so all the calculated results may be considered as real characteristics of the tube in a stress-deformed state that proves the strength safety of the dimpled tube for the doubled operating pressure.

Field Trial System

The Field Trial System (FTS) comprises of two flow ducts bypassing the existing economizer from both sides as shown in Figure 8. Figure 11 illustrates a finned tube branch of the FTS. Each duct was equipped with an adjustable deflector, control damper and turning vanes to set and maintain the equal flow rates in both test sections.

Figure 11. Overall views of the test ductwork (finned tubes)

Three test sections were fabricated for the field trial - finned tube (industry standard), dimpled tube (in-line arrangement) and dimpled tube (staggered arrangement). Figure 12 illustrates the inner view of the finned and dimpled test sections.

a b

Figure 12. Test sections: a – finned, b – dimpled (staggered)

Mass flow meters were installed at the flow outlets while temperature sensors were installed at the flow inlet and outlet. Pressure drop across the test section will be measured by differential pressure gauge. Some measurement equipment was installed at the existing economizer to monitor the baseline unit operation. The measurement system layout is presented in Figure 13.

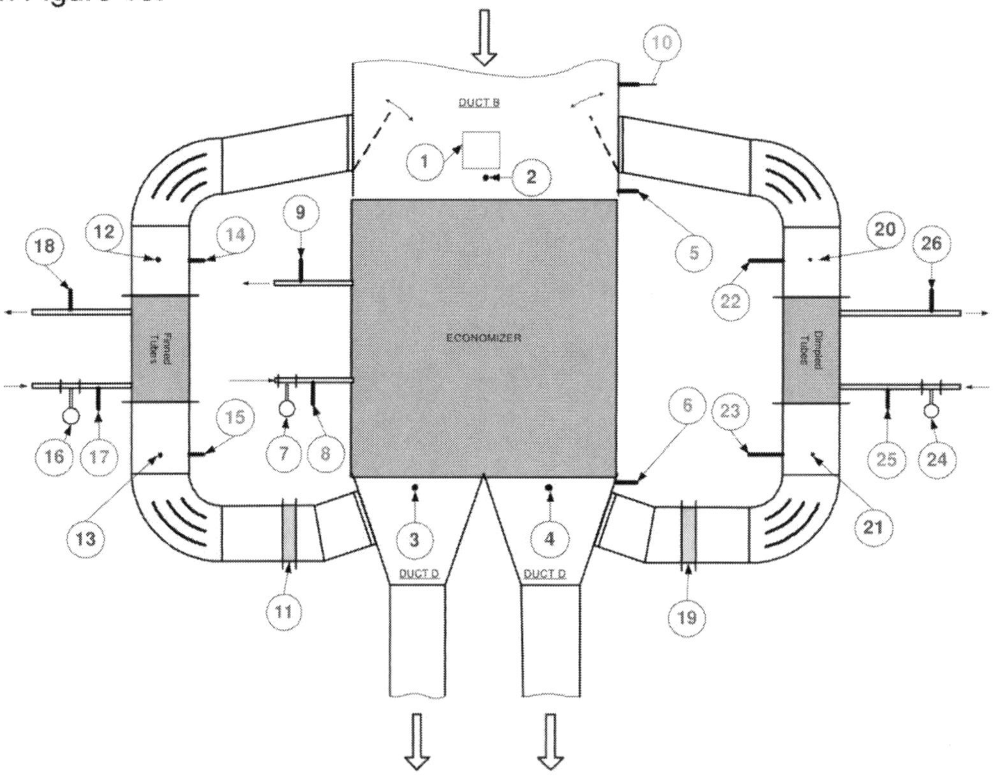

Figure 13. Measurement system layout:
Flow measurement: 1, 7, 11, 19, 16, 24
Temperature measurement: 2-4, 9, 12-13, 18, 20-21, 26
Pressure measurement: 5-6, 8, 14, 15, 17, 22-23, 25
Flue gas composition: 10

The following experimental conditions are expected for the comparative data collection:

- flue gas flow rate for each test section: 3850 CFH (3% of the total flow rate)
- flue gas inlet temperature at the economizer and test sections: 590°F
- flue gas outlet temperature:
    - economizer: 430°F
    - test sections: 370°F -380°F
- flue gas pressure drop across the economizer: 1.4 in WC
- water mass flow rate for each test section: 8 GPM
- water inlet/outlet temperature at the test section: 60°F/155°F

Conclusions

The preliminary analysis demonstrates the strong potential for significant savings in energy (fuel), capital costs and increase in throughput by manufacturing or retrofitting the convective section for fired process heater with dimpled tubes as opposed to finned and shock tubes. Based on the experimental data the following benefits are estimated for the industrial application based on a natural gas price of $5/MMBtu and a domestic US population of 700 high temperature fired heaters:

- energy savings per each 1% of fuel use decrease = $250-350K per heater annually
- reduced expenses per each 1% retrofit of installed base = $50-100K annually
- gain per each 1% increase in throughput = 10,000 tons/heater of product (assuming $500/ton of ethylene ~ $5M)

Currently GTI along with the Industrial Partners and Vendors are running the field trial at the selected Chemical Industry site. The preliminary field trial results are expected in the second quarter of 2005.

Based on fundamental research results and understanding of the vortex flow formation mechanisms it is also expected significant reduction in fouling rate resulted in non-increase of stack temperature as well as lowering the operating costs due to reduced maintenance, cleaning, etc. In the framework of the field trial a 6-12 months monitoring is planned to prove the favorable fouling performance.

Acknowledgments

The authors are grateful to US Department of Energy (award DE-FC36-01ID14089), Gas Research Institute and Members of GTI's Sustaining Membership program for their financial support to this work as well as Industrial Partners for valuable assistance in design review and results analysis.

References

1. A.Schukin, A.Kozlov, Y.Chudnovsky, R.Agachev. "Intensification of Heat Exchange by Spherical Depressions. A Survey", Applied Energy: Journal of Fuel, Power and Heat Systems, vol. 36, no. 3, 1998.

2. I.Shrader, A.Daschchyan, and M.Gotovskii. "High-Intensity Tubular Air Heaters". Thermal Engineering, vol. 46, no. 9, 1999.

3. Ya. Chudnovsky. "Heat Transfer and Fluid Flow in a Spherical Cavity," Paper presented at Second European Thermal Sciences and UIT National Heat Transfer Conference, Rome, Italy, May 1996.

4. S.Isaev, A.Leontiev, P.Baranov and A.Usachev. "Bifurcation of Vortex Turbulent Flow and Intensification of Heat Transfer in a Hollow", Doclady Physics, vol. 45, no.8, 2000.

5. Ya. Chudnovsky, H.Kurek, A.Kozlov. "Dimpled Tube Technology For Heat Transfer Enhancement In Chemical Industry Process Heaters". Proceedings of the 2nd Natural Gas Technologies Conference, Phoenix, AZ, February 2004.

6. Ya.Chudnovsky, H.Kurek. "Vortex Heat Transfer Enhancement For Waste Heat Recovery In Chemical Industry Process Heaters". Proceedings of the AIChE Spring National Meeting and DOE ITP Peer Review, New Orleans, LA, April 2004

7. Ya.Chudnovsky, A.Kozlov. "Development and Field Trial of Dimpled-Tube Technology for Chemical Industry Process Heaters". Technical Report GRI-05/0052, March 2005.

1e Experimental Methods Defining the Trickle-pulse Transition in Two-phase Flows Through Fixed Beds

John R Torczynski, Lindsey R Evans, Marcia A Cooper, Raymond O Cote, Timothy J O'Hern

Methods to experimentally determine the conditions at which downward co-current gas-liquid flows through a fixed bed of particles confined within a cylindrical column transition from trickle flow to pulse flow are investigated. The pulse-flow hydrodynamic regime is a popular process intensification strategy for many petrochemical applications because of the increased mass and heat transfer rates, increased throughput without flooding, and complete wettability of the bed particles compared to the trickle-flow regime. Knowing the operating regime and, in particular, the conditions leading to trickle-pulse transition in an industrial application is of critical importance in order to exploit the benefits gained from such intensification strategies. A quantitative measure of this important regime transition has not been previously established. Visual observation through a clear column is the usual technique to determine transition in the lab but is an impractical diagnostic for commercial applications in the field as well as being subjective. Additionally, the method of visual observation becomes increasingly difficult in beds of spherical particles as the particle diameter d becomes comparable to the column diameter D. In this case, the increase in void fraction near the inner column surface causes much of the liquid to flow down the wall, obstructing the view of the bulk flow within the bed. Experiments are carried out in a column with an inner diameter of 9.5 cm and a length of 2.4 m. Co-current streams of air and water flow downward through the packed bed, which consists of polished, solid, spherical, monodisperse particles ("beads") with mean diameter between 0.64 cm and 1.27 cm (D/d between 7.5 and 15). Measurements from pressure transducers and conductance probes are used to determine the conditions under which the trickle-pulse transition occurs over a range of gas and liquid flow rates and particle diameters. These data generate a more robust definition of the trickle-pulse transition that does not currently exist in the literature. The financial support of the US Department of Energy Office of Industrial Technologies, Brian Valentine, contract monitor is acknowledged. Sandia is a multiprogram laboratory operated by Sandia Corporation, a Lockheed Martin Company, for the United States Department of Energy's National Nuclear Security Administration under Contract DE-AC04-94AL8500.

1f Discussion Session

Roshan Jachuck

Discussion on accelerating the deployment of novel reaction technology

2a Keynote Presentation

Roshan Jachuck

The state of the art in BioIntensification

2b Intensified Liquid Phase Reactions in Tethered Catalyst Microchannel Reactors

John Brophy and Kai Jarosch

Velocys is developing continuous liquid phase microchannel reactors incorporating 'homogeneous' catalyst tethered to the reactor wall. This novel approach improves many aspects of liquid phase chemical processing including safety, productivity and costs.

Liquid phase continuous flow reactors are potentially cleaner, safer and more efficient than batch chemical production. Continuous flow reactors offer better control of temperature and residence time; thus, reducing inventories of hazardous materials and improving overall operating and capital costs. Microchannel reactors accentuate the continuous flow benefits by improving both heat and mass transfer. These features enable close control of reaction conditions; thereby, improving selectivity and maximizing yield.

The concept of performing liquid phase reactions in tethered catalyst, microchannel reactors has been demonstrated for two base catalyzed reactions:

1) Knoevenagel Reaction - an important carbon-carbon bond forming reaction involving a base catalyzed condensation of aldehyde and activated methylene

2) Michael Reaction - the conjugate addition of nitroalkanes to a,b-unsaturated carbonyl compounds.

Compared to the industry standard - batch, stirred tank reactors - the Velocys' tethered catalyst approach has the potential for many salient advantages, including:

- High product yield and selectivity - Tightly controlled mixing, short diffusion distances and laminar flow regimes allow the reaction to progress uniformly, producing higher purity compounds and fewer unwanted by-products.

- Eliminating the need for catalyst recovery - Tethering catalyst in the reactor permits the reactions to proceed in the controlled environment inside the reactor, while preventing the need to recover the valuable catalyst material downstream.

- Rapid, in-line heating and cooling - The improved heat transfer capabilities of microchannel reactors allow precise heat application or removal to achieve optimal reaction conditions.

- Modular approach - Incremental capacity can be added or removed by increasing or decreasing the number of fixed capacity reactors. Also, sequential arrangements are possible to continuously process complicated chemicals.

- Small, inexpensive reactors - Fast reaction kinetics enabled by microchannel architecture allow small reactors to produce commercially significant quantities of chemicals.

This paper will discuss the results for both reactions above and compare these with similar homogenous and heterogeneous catalysts in batch systems. The results show consistently higher rates of conversion with no loss of catalyst activity. Also included is discussion of the benefits of this novel reactor system for high productivity, continuous processing.

2c Gold Catalyzed Cyclohexane Oxidation

Mikhail I. Khramov

The use of gold catalysts for cyclohexane oxidation promises significant improvements over existing technology. Already, CO has been successfully oxidized into CO2 with gold as a catalyst, and investigations have shown that gold nanoparticles are highly efficient catalysts for the oxidation of cyclohexane. The use of gold catalysts for cyclohexane oxidation could improve conversion rates and yields, and provide the chemical industry with significant potential energy savings, while at the same time reducing waste and the use of raw materials. This paper will discuss research results from the study of supported gold catalysts for cyclohexane oxidation.

2d Catalytic Reactions for the Production of Biomarkers from the Biological Warfare Agent Anthrax

Aaron N Nackos, Calvin H Bartholomew, Edgar D Lee, Milton L Lee, Phillip R Smith, zhijun Jia

Spores of Bacillus anthracis, or anthrax, are highly toxic to humans in only nanogram quantities. They can be and have been employed as deadly biological weapons. Portable analyzers that can rapidly detect and identify weaponized biological agents such as anthrax spores are critically needed. Semi-portable pyrolysis/MS has been used in their detection. A potential advance in detection technology is possible through use of heterogeneous catalysts for the production of biomarkers. Dipico-linic acid and fatty acids are important biomarkers for detection of anthrax spores. In the present work, a superacid catalyst of the Keggin-structure, H3PW12O40, was observed to have high activity for production of the (di)-methyl ester of dipicolinic acid (DPA). Ths catalyst is active for the esterification reaction at room temperature. The complete conversion of DPA to (di-)methyl ester occurs at 25oC in less than 49 hours; 90% conversion is reached at 95oC in 2 minutes. The Similar results are also observed for esterification of fatty acids (C12 to C18).

2e Presentation by Dr. Laurence Weatherly

Laurence Weatherly

Presentation by Dr. Laurence Weatherly

2f Discussion Session

Roshan Jachuck

Discussion on accelerating the deployment of bio and process intensification via innovative chemistry.

3a Keynote Address on Novel Separation Technology

Lawrence Weatherly

Electrostatic fields can play a major role in enhancing liquid/liquid and gas/liquid separation duties. This presentation will provide an overview of the opportunity offered by the use of electrostatic fields for achieving energy efficient separation for the chemicals sector.

3b Removal of Heavy metal ions by Magnetically Modified Zeolite

Choong Jeon, In Wook Nah, Jae Ho Kwak, Kyung-Yub Hwang

A simple route for decorating zeolite with iron oxide nanoparticle using urethane was described. The prepared Magnetically Modified Zeolite (MMZ), characterized by XRD, SEM, EDS, and pore size distribution, was found to retain the parent zeolite. The removal of heavy metal ions from diluted solutions by sorption onto MMZ, magnetic separation, and regeneration were studied at laboratory scale. Effects of initial heavy metal ions such as Pb, Cd, Cu, Zn ion concentrations on removal efficiency, adsorption capacity of the MMZ, and adsorption isotherms were investigated.

3c A Novel, Highly Efficient and Economic Purification Process Revolutionizing PTA Production

Costantino Braggiato, Randi Wytcherley, Ron Gualy, Tai-Li Chou

GTC Technology, with support from the Department of Energy Chemicals Subprogram of the Industrial Technologies Program, is investigating a bench scale proven process for the purification of terephthalic acid. The new process significantly reduces enery use, eliminates use of hazardous chemicals, and produces polyester grade material from lower purity, lower cost para-xylene feedstocks.

Production of Magnetic Beads for bio-separation – Process scale-up

Banert, T., Peuker, U. A., Clausthal University of Technology, ICVT, Leibnizstraße 17, 38678 Clausthal-Zellerfeld, Germany, phone: ++49 5323 722187, e-mail: urs.peuker@tu-clausthal.de

Abstract

A modified solution process is used to produce super-paramagnetic nano composite powder containing functional groups for adsorption application. The powder is produced in a spray tower. The feed is a polymer solution containing two fractions of different particles: magnetite (14 nm) and polymer ion exchanger (150 nm). The spray dried magnetic beads are submitted to different characterisation methods: confocal Laser Scanning Microscopy (LSM), Small Angle X-Ray Scattering (SAXS), Alternating Gradient Magnetometry (AGM).

Introduction

Magnetic Separation seems to be a promising unit operation for down stream processing of bio suspensions [29][30][31][32][33]. The technology consists of two main technology carriers: the magnetic separator and the magnetic bead [36]-[52]. The magnetic bead is in most cases a composite material with a polymer matrix [2][3][4][5][6][7][10], in which magnetic particles are incorporated. The surface of the magnetic beads can be modified to adsorb the substance [8][9][11][12][13][14][15], which has to be recovered.

The separation technology using magnetic beads is widely-used in the micro scale. Complicated separations of proteins or DNA-fragments are operated in biological laboratories. But the scale up of this principle to use the magnetic separation technology in production scale is limited due to the high costs of the beads.

Magnetic beads have a diameter of 100 nm to about 100 µm. The separation effect derives from the sorption, complex formation or covalent bonding of the target substance to the bead surface. With increasing accessible surface the possible load per volume unit of a magnetic bead also increases.

The common synthesis process uses emulsion or suspension polymerisation [1][8][15]. The magnetic properties are provided by fine magnetic particles, e. g. magnetite. In presence of dispersed magnetic particles the polymerisation reaction is started. The polymer incorporates the magnetic particles into its matrix. The functionalization of the surface is prepared in further process steps. The presence of iron, which is due to the magnetic filler, as a catalyst negatively, influences the functionalization reactions. The polymerisation processes also show some weakness in the quality of dispersion of the magnetic filler [1]. This is of relevance, when super-paramagnetic properties of the beads are requested. The recovering of the beads from a magnetic separator is less complicated, in case of a magnetisation without magnetic remanence [30][26].

For scale up purposes the strategy of the bead production has been strongly modified. The main target is to set up a process, which is able to produce at least some kg per hour of magnetic beads with a basic functionalization. The requested properties of the beads are: super-paramagnetism, sufficient filler content for separation, sufficient surface-volume ratio and accessible functionalised surface groups.

The solution process

The solution process bases on a polymer solution to manufacture polymer particles. The solidification mainly is made by spray-drying [20][21][22][23][24][25] an alternative is precipitation using an anti-solvent. This method can be used to produce nano-composite bulk materials [17][18][19].

The main innovation can be seen in the process scheme (Figure 1). The surface functionalization and the magnetic properties are first developed independently. The surface function to interact with proteins is provided by nano-scale polymer particles synthesised in emulsion polymerisation. The size is about 80 to 250 nm. After synthesis a second reaction to provide the functional surface groups is conducted. The basic surface functionalization is provided by amino or sulfone groups.

On the other side a magnetic fluid is produced, which bases on a solvent, in which the future matrix polymer is soluble.

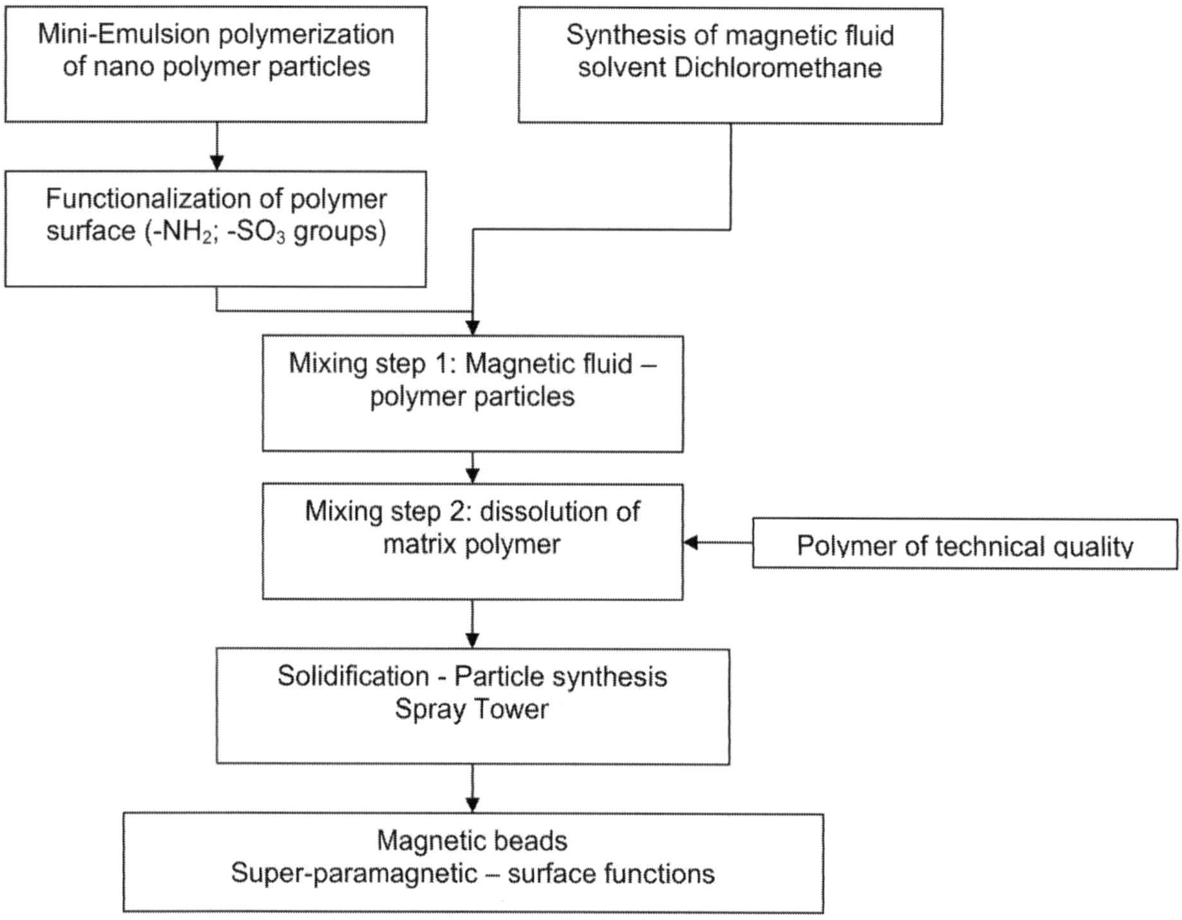

Figure 1 Process scheme

The process starts from the magnetic fluid. Dichloromethane is used as mother liquid of the magnetic fluid. Oleic acid is used as stabilising agent. The synthesised nano scale ion exchangers are dispersed into the magnetic fluid. The matrix polymer is dissolved as well into the magnetic fluid.

The dispersion containing super-paramagnetic magnetite particles (< 20 nm), nano-scale ion exchange latex particles and dissolved polymer is pumped into a spray tower. Using a two component jet the fluid is dispersed into droplets of 5-15 micron, which results in a particle size for the magnetic beads of 2-10 micron.

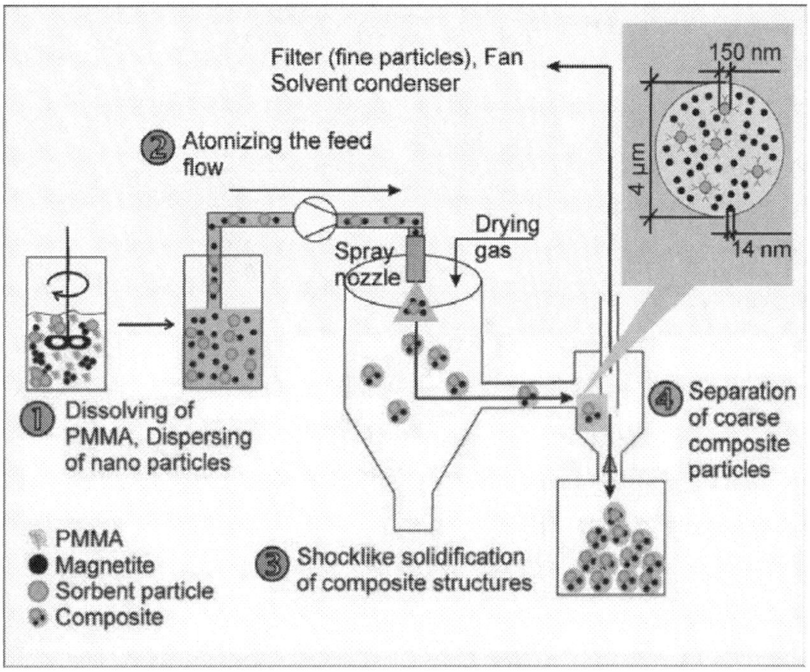

Figure 2 Process design of the solution process using a spray tower

Material & methods

PMMA granulate material is of technical quality and purchased from Rhenus AG & Co. Oleic acid, Iron(II)sulfate Heptahydrate ($FeSO_4*7H_2O$) and Iron(III)chloride Hexahydrate ($FeCl_3*6H_2O$), Dichloromethane (DCM), Ammonium hydroxide (26 % NH_3) and Petrolether are purchased from Fluka AG and used as received.

According to [26][28] the magnetite nanoparticles are produced by co-precipitation from an aqueous solution of Iron(III)chloride Hexahydrate and Iron(II)sulfate Heptahydrate with a molar ratio of Fe^{3+}/Fe^{2+} = 2:1. The iron solution is precipitated with concentrated ammonium hydroxide in excess. The precipitation process is operated under vigour stirring in a glass beaker. Alternatively, in respect of scale up possibilities, an ultrasound flow through reactor has been used [27]. By default 24.05 g $FeSO_4$* $7H_2O$ and 47 g $FeCl_3$* $6H_2O$ are dissolved in 1000 ml distilled water and heated to 70 °C. Then, 60 ml ammonium hydroxide is added rapidly. Deep black magnetite precipitates at once. The magnetite yield is 20 g/L. The suspension has been stirred for yet another 30 minutes at 70°C in order to evaporate ammonium salts. Otherwise, the precipitation salts adsorbe onto the particle surface and the transfer into the organic phase is obstructed. The organic phase contains 0.1 to 1 $g/g_{Magnetite}$ of Oleic acid as detergent. Finally after the transfer step a stable oil-based ferrofluid with a yield of 20 $g_{mag.}/l_{DCM}$ is obtained. Particle size is measured by PCS (Zetasizer Nano ZS, Malvern Inc.) to control precipitation and stabilization.

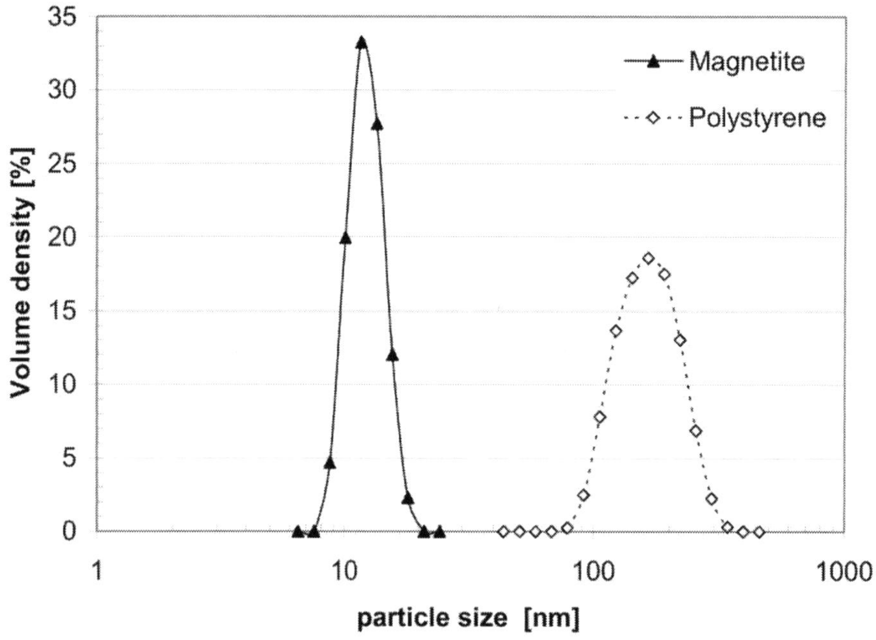

Figure 3 Particle size distribution of the dispersed phases: magnetite and latex (polystyrene) particles

The synthesis of the nano-scale polymer particles starts with the mixing of 6 g Styrene and 2.2 g Divinylbenzene (DVB). The mixture is added to a surfactant solution consisting of 24 g distilled water and 72 mg Sodium Dodecyl Sulphate (SDS). The first step of emulsifying includes 30 min of vigour stirring to come to a macro emulsion. In the second step after 2 min of sonification (US 100 W at 20 kHz) the droplets are further broken up to come to a mini-emulsion. To start the polymerisation 20 mg of Sodium Peroxidisulfate is added and the temperature is increased up to 70°C. The polymer is transferred directly from aqueous to non-aqueous phase (DCM) and the particle size distribution can be measured (Figure 3). Depending on the functionalization the further reactions are chosen, e.g. Sulfonation using Chlorosulfonic acid.

As a second monomer Vinyl Benzyl Chloride (VBC) has been used. In this case the monomer and SDS concentration has to be lower in mini-emulsion, otherwise strong agglomeration occurs. The functionalization of die poly-VBC uses Trimethylamine.

After suspending and de-agglomeration of the functionalised polymer particles in the magnetic fluid the polymer PMMA is directly dissolved into the particle fluid, which bases on Dichloromethane. The viscosity of the suspension does not exceed 50 mPaS and is still low enough to ensure perfect mixing of magnetite and polymer in the molecular scale. The mass of dissolved polymer depends on the expected filling degree. The composite material is solidified in a conventional spray dryer (Buechi B-191): The feed flow of the spray dryer consists of dissolved polymer, dispersed sterically stabilized magnetite and dispersed nano-scale ion exchange particles. The atomizer produces a fine spray with a small droplet size of about 20 micron, which is dried in the spray tower at 40 °C. The continuous organic and the dispersed organic and inorganic phase do not separate during the drying process and solidification. The magnetite is finely dispersed within the PMMA-matrix. The product is a highly filled polymer-nano-composite-powder with a mean particle size of 2 to 10 micron. Subsequently the nano-compound is washed several times to remove a film consisting of excess Oleic acid and non-encapsulated magnetite.

The content of magnetite in the ferrofluid, in the unwashed and in the washed particles is quantified with a spectrophotometer (Hach Company, DR 4000) using the 1.10 Phenantroline method.

Small angle X- ray scattering (SAXS, CuK_α radiation) profiles have been recorded to determine the mean particle size of magnetite within the composite material. AGM (Alternating Gradient Magnetometry) measurements are carried out at room temperature to get information about the saturation magnetization and remanence of the composite powder.

Results

Confocal Laser Scanning Microscopy clearly shows the distribution of the nano scale ion exchange particles within the magnetic bead.

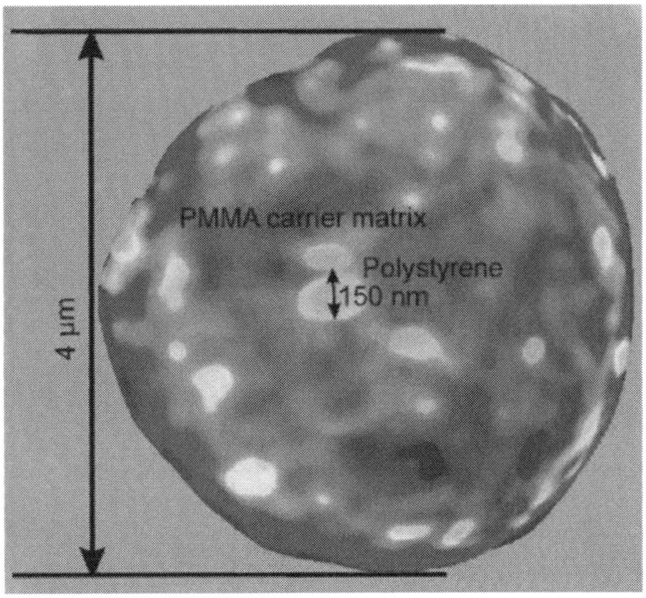

Figure 4 Overlay picture (reflection – fluorescence) made by a confocal laser scanning microscope. Filler material: non functionalized PS particles by mini-emulsion polymerisation.

Two pictures of the same bead have been recorded, one in reflection mode the second in fluorescence mode. The confocal LSM has a maximal resolution of about 100 nm. Therefore the nano magnetite is invisible with this technique. The two polymers (PMMA shell, PS filler) have different optical properties. In the reflection mode the PS filler particles can be detected. In the fluorescence mode the surface morphology of the bead itself is recorded. The overlay of both pictures finally shows the position of the PS nano particles within the magnetic bead (Figure 4). The size is calculated form the resolution and number of pixels. The size fits well to the size measured after mini-emulsion polymerisation (Figure 3).

The magnetic properties have been quantified with AGM measurements (Figure 5). The saturation magnetization of the composite material is reached with the magnetic field strength of about 10^6 A/m. The measurements are exploited as specific magnetization per mass composite (straight line) and per mass magnetite within the composite (dotted line). The magnetization of the composite material amounts to 25.59 Am^2/kg at a magnetite content of 30 mass%. If the magnetization is referred to the mass of magnetite itself the magnetization amounts to 85.3 Am^2/kg, which is quite close to the literature [26]. This shows that the artificial magnetite is of excellent quality and the encapsulation process does not change the magnetite.

A close look at low magnetic field strength shows that the curves hit the ordinate at very low magnetization. The remanence of the composite material amounts to 0,574 Am^2/kg. Thus the ratio of saturation magnetization and remanence is 0.02. This is a quite good value, it can be stated that the super-paramagnetic properties have been incorporated within the composite material. The high filling degree provides large magnetic forces during magnetic separation processes.

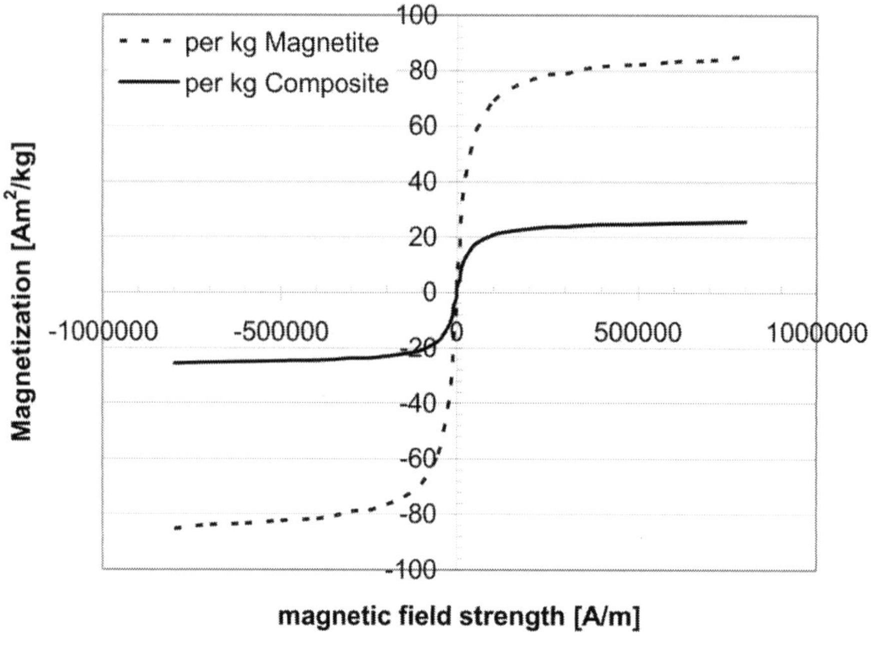

Figure 5 Magnetic properties auf the composite material

Another question of the synthesis process is the question of encapsulation efficiency. The magnetic beads have to be washed after the spray process because the excess detergent Oleic acid is not incorporated into the composite material. It forms a kind of surface film, which can be removed by intense washing. The washing media has a dark black colour after use. The analysis shows that with the Oleic acid film some of the magnetite is removed from the magnetic beads. The more excess Oleic acid is contained in the feed dispersion / solution, the thicker the surface film is. Experiments have been made to determine the optimal Oleic acid content (Figure 6). The mass ratio of Oleic acid to polymer has been varied from 0.3 $g_{Oleic\ Acid}/g_{PMMA}$ up to 1 $g_{Oleic\ Acid}/g_{PMMA}$. The minimum value of 0.3 $g_{Oleic\ Acid}/g_{PMMA}$ is due to the minimal quantity of Oleic acid necessary to stabilize the ferrofluid at maximum magnetite content of the composite material.

The encapsulation efficiency decreases with increasing content of Oleic acid. It also slightly decreases with increasing magnetite content. Regarding the composite formation it can be stated that the optimal concentration of detergent in the dispersion is equal to that adsorbed at the magnetite surface, which ensures the stabilization of the ferrofluid. All additional Oleic acid interferes with the encapsulation of the magnetite. Therefore the further processing has already to be kept in mind during the preparation of the ferrofluid. The encapsulation efficiency amounts to about 95% at the lowest detergent concentration of 0.3 $g_{Oleic\ Acid}/g_{PMMA}$ and at magnetite content of 50 mass%. This shows that it is easily possible to synthesize highly filled nano composite materials with the spray process.

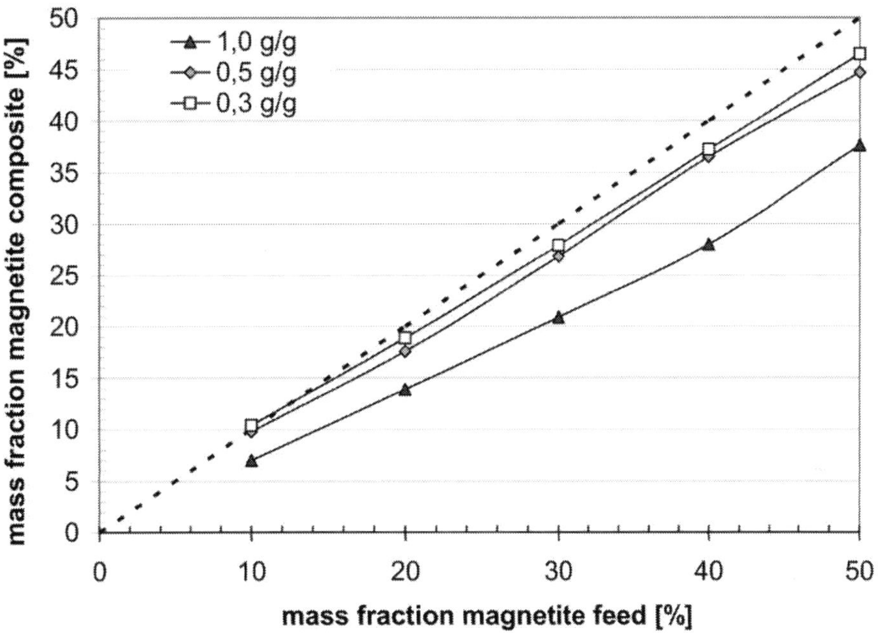

Figure 6 Encapsulation of magnetite

The Small Angle X-Ray Scattering (SAXS) can be plotted in a so called Guinier-Plot (Figure 7). From the slope of the curve, logarithm of intensity over square of scattering angle (2 θ), the characteristic length the composite material can be calculated. The measurements are made with a spray dried powder containing 22.8 mass% of magnetite, but no functionalized polymer particles. One reason for the measurements using a homogeneous composite material is that the presence of the non ion exchange particles leads definitely to magnetite free areas within the composite structure. Therefore a change in the characteristic length has to be expected. The calculations give a mean particle size of magnetite within the composite material of 14.5 nm. This excellently corresponds to the particle size distribution measured in the ferrofluid (Figure 3).

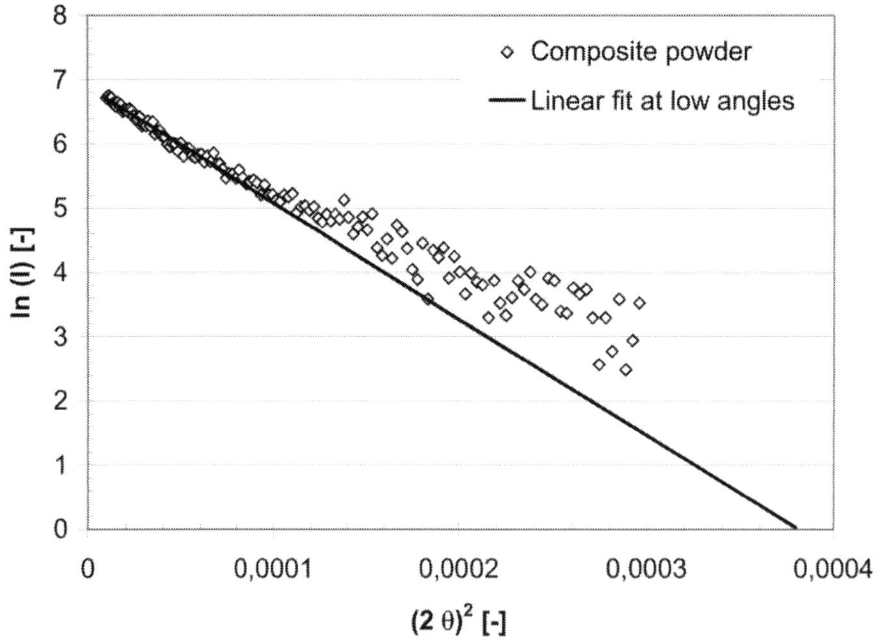

Figure 7 SAXS-Measurement: mean particle size within the composite material

The ion exchange capacity of magnetic beads has been quantified. The magnetic beads containing aminated nano ion exchange particles are loaded with Chloride ions in Hydrochloric acid (5%). After ion exchange using Sodium sulfate the exchange capacity of the magnetic beads is calculated to 0,47 $mVal/g_{beads}$.

Application / Process integration

The possible process integration is shown in Figure 8. Due to the engineering approach of the synthesis the beads are available in larger quantities than today. Therefore some kilograms of beads can be added to e.g. a fermentation broth. During suspension the (selective) adsorption of the main product at the functional areas of the beads occurs. In the subsequent step the beads can be removed with a technical magnetic separator (HGMS, HGF) [34][35]. The large advantage of this separation process in technical scale is the lack of any filtration steps. After a short washing step the beads are transferred to the elution step, where the main product is removed from the beads. This is done e.g. by pH-shift and desorption. Finally the beads and the product are separated. The beads are separated (magnetically, conventionally) from the elution buffer and can be retransferred to the process and the sorption starts again.

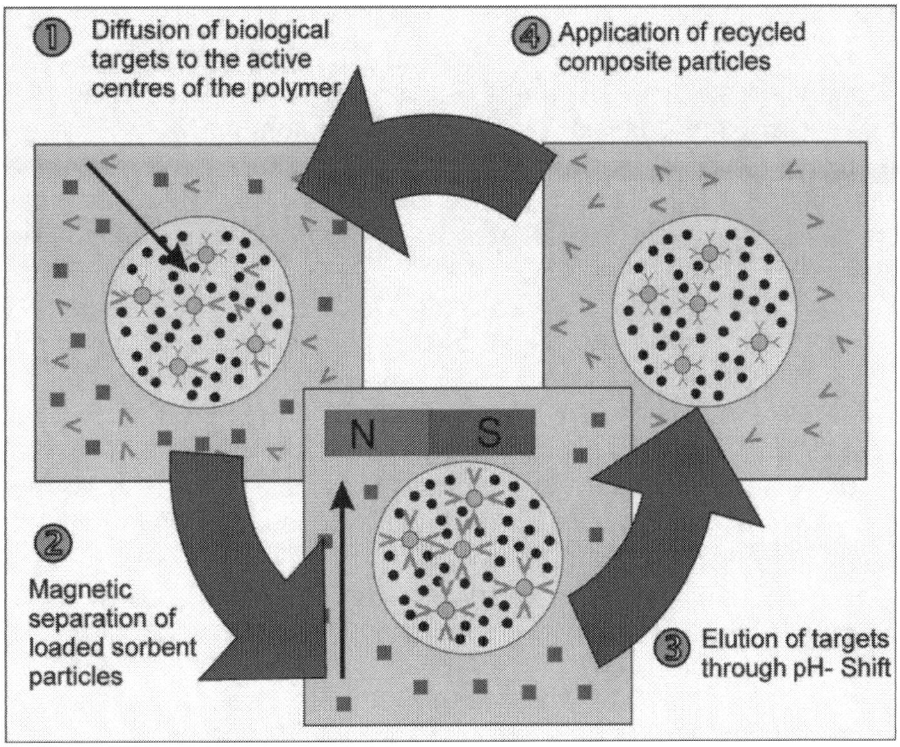

Figure 8 Scheme of large scale process design using magnetic beads.

Conclusions

A new engineering approach has been shown, which allows the production of large quantities of functionalized magnetic beads. The sorption properties are provided by nano scale ion exchange particles, which are synthesised in a separate process. In a final step the nano ion exchangers are incorporated into the magnetic beads. All processes applied (precipitation, mini emulsion polymerisation, mixing, spray drying) for the synthesis can be scaled up for the production in kilogram or even ton scale.

The experimental results show super-paramagnetic properties of the composite particles, which is due to the use as magnetic filler and the excellent dispersion of nano scale magnetite particles. The incorporation of the nano ion exchange particles within the bead and their accessibility has been made visible with confocal LSM.

With this synthesis process of magnetic beads the application of selective sorption and magnetic separation in the field of the so called white biotechnology becomes less expensive and with it more probable.

Acknowledgements

We like to thank our co-operation partners: Prof. Opperman, Dr. Langhoff (Clausthal), Dr. Franzreb (Karlsruhe).

Literature

[1] Ramirez, L., Landfester, K., magnetic Polystyrene Nanoparticles with high Magnetite Content Obtained by miniemulsion Processes, Macrommol. Chem. Phys., Vol. 204, 2003, pp. 22-31.

[2] Phanapavudhikul, P. et.al., Design and Performance of Magnetic Composite Particles for Separation of Heavy Metals from Water, J. of Environmental Science and Health, Vol. A38, 2003, pp. 2277-2285.

[3] Huang, Z., Tang, F., Preparation, structure, and magnetic properties of polystyrene coated by. Fe_3O_4 nanoparticles, Journal of Colloid and interface Science, Vol. 275, 2004, pp. 142-147

[4] Lu, S., Preparation of Molecularly Imprinted Fe3O4/P(St-DVB) Composite Beads with magnetic susceptibility..., J. of applied polymer science, Vol. 89, 2003, pp. 3790-3796.

[5] Cocker, T. M., et.al., Preparation of Magnetically Susceptible Polyacrylamide/Magnetite Beads for Use in Magnetically Stabilized Fluidized Bed Chromatography, Biotechnology and Bioengineering, Vol. 53, 1996, pp. 79-87.

[6] Khng, H. P., et.al., The Synthesis of Sub-Micron Magnetic Particles and Their Use for Preparative Purifikation of Proteins, Biotechnology and Bioengineering, Vol. 60, 1998, pp. 419-424.

[7] Marik, J., et.al., Magnetization of large polystyrene peptide beads for capturing and expandig cancer cells, J. of Magnetism and Magnetic Materials, Vol. 264, 2003, pp. 153-157.

[8] Bergemann, C. et.al., Magnetic ion-exchange nano- and microparticles for medical biochemical and molecular biological applications, J. of Magnetism and Magnetic Materials, Vol. 194, 1999, pp. 45-52.

[9] Xie, X., et.al., Preparation and application of surface-coated superparamagnetic nanobeads in the isolation of genomic DNA, J. of Magnetism and Magnetic Materials, Vol. 277, 2004, pp. 16-23.

[10] Oster, J. et.al., Polyvinyl-alcohol-based magnetic beads for rapid and efficient separation of specific or unspecific nucleic acid sequences, J. of Magnetism and Magnetic Materials, Vol. 225, 2001, pp. 145-150.

[11] Uhlen, M., Magnetic Separation of DNA, Nature, Vol. 340, 1989, pp. 733-734.

[12] Ennis, M., Wisdom, B., A Magnetizable, Solid Phase for Enzyme Extraction, Applied Biochemistry and Biotechnology, Vol. 30, 1991, 155-164.

[13] An, X., et.al., Preparation of highly magnetic chitosan particles and their use for affinity purification of enzymes. J. Chem. Technol. Biotechnol., Vol. 78, 2003, pp. 596-600.

[14] Caruso, F., et.al., Nanoengineering of Inorganic and Hybrid Hollow Spheres by Colloid Templating, Science, Vol. 282, pp. 1111-1114.

[15] Kietzke, Thomas, Neher, Dieter, Landfester, Katharina, Montenegro, Rivelino, Novel approaches to polymer blends based on polymer nanoparticles, Nature Materials, Vol. 2, 2003, pp. 408-412

[16] Tyc, I., Green, B. R., Phenol-Formaldehyde Based Weak-Base Resins for the Recovery of Gold, Solvent extraction and ion exchange, Vol. 12, 1994, pp. 817-830.

[17] Banert, T. Peuker, U. A., Herstellung von hochgefüllten Fe_3O_4/PMMA-Nanokompositen im Sprühprozess, Chem. Ing. Tech., Vol. 77, 2005, pp. 224 - 227

[18] Zhou,X.D., Zhang,S.C., Effects of the solvent on the particle morphology of spray dried PMMA, J. of Materials Science, Vol. 36, 2001, pp. 3759-3768

[19] Zhou,X.D., Gu,H.C., Synthesis of PMMA-ceramics nanocomposites by spray process, J.of Materials Science, Vol. 21, 2002, pp. 577-580
[20] Arimura, Hidetoshi, Ohya, Yuichi, Preparation of Polylactide Microspheres Having Positively or Negatively Charged Surfaces, Macromol.Biosci, Vol. 3, 2003, pp. 18-25
[21] Bain, D.F., Munday, D.L., Thermal studies on spray-dried biodegradable microspheres from various organic solvents, Journal of Pharmacy and Pharmacology, Vol. 50, 1998, pp. 161
[22] Bain, D.F., Munday, D.L., Solvent influence on spray-dried biodegradable microspheres, Journal of Microencapsulation , Vol. 16, 1999, pp. 453-474
[23] Guo,X., Bodmeier,R., Polymeric microparticles prepared by spray-drying a drug-containing,aqueous colloidal acrylic polymer dispersion,Eudragit RS30D, S.T.P.Pharma Sciences, Vol. 6, 1997, pp. 521-528
[24] Millqvist-Fureby, Anna , Malmsten,Martin, An Aqueous Polymer Two-Phase System as Carrier in the Spray-Drying of Biological Material, J.of Colloid and Interface Science, Vol. 225, 2000, pp. 54-61
[25] Göktürk, Halit S., Fiske, Thomas J., Effects of particle shape and size distributions on the electrical and magnetic properties of Nickel/Polyethylene composites, Journal of Applied Polymer Science, Vol. 50, 1993, pp. 1891-1901
[26] Schwertmann, U., Iron oxides in the laboratory : preparation and characterization, Weinheim Wiley-VCH, 2000
[27] Banert, T., et.al., Kontinuierliche Fällung im Ultraschall-Durchflussreaktor am Beispiel von Eisen-(II,III)-Oxid, Chem. Ing. Tech., Vol. 76, 2004, pp. 1380-1381
[28] Hoffmann, Chr., Einsatz magnetischer Separationsverfahren zur biotechnologischen Produktaufarbeitung, Dissertation Universität Karlsruhe, 2002
[29] Hubbuch, J. J., et.al., High Gradient magnetic separation versus expandes bed adsorption: a first principle comparison, Bioseparation, Vol. 10, 2001, pp. 99-112.
[30] Heeboll-Nielsen, A., et.al., Superparamagnetic Adsorbents for High-Gradient Magnetic Fishing of Lectins Out of Legume Extracts, Biotechnology and Bioengineering, Vol. 79, 2002, pp. 301-313.
[31] Langer, R., Peppas, N. A., Advances in Biomaterials, Drug Delivery, and Bionanotechnology, AIChE Journal, Vol. 49, 2003, 2990-3006
[32] Hubbuch, J. J.,Thomas, O., High-Gradient Magnetic Affinity Separation of Trypsin from Porcine Pancreatin, Biotechnology and Bioengineering, Vol. 87, 2004, pp. 311-323.
[33] Zhang, Zhanren, O'Sullivan, Deirdre A., Magnetically stabilised fluidised bed adsorption:Practical benefit of uncoupling bed expansion from fluid velocities in the purification of a recombinant protein from Escherichia coli, Journal of Chem. Tech.and Biotech., Vol. 74, 1999, pp. 270-274
[34] Kohlhaupt U.,Franzreb, M., Separation of finest particles by high gradient magnetic filtration, Proc. Partec Nürnberg 2004
[35] Meyer, A.; et.al.; Demonstration of a Strategy for Product Purification by High-Gradient Magnetic Fishing: Recovery of Superoxide Dismutase from Unconditioned Whey, Biotechnology progress, Vol. 21, 2005, pp. 244-255
[36] Bidan, Gerard,Material based on an electronic conductive polymer incorporating magnetic particles and its production process, US 5,264,157 A,23.11.1993
[37] Bradbury, David, Magnetic particles, a method for the preparation thereof and their use in the purification of solutions, US 5,855,790 A,05.01.1999

[38] Charmot, Dominique, Magnetizable composite microspheres of hydrophobic crosslinked polymer,process for preparing them and their application in biology, US 5,356,713 A,18.10.1994
[39] Elaissari, Abdelhamid, Magnetic particles, method for obtaining same and uses for separating molecules, US 6,521,341 B1,18.02.2003
[40] Lange, Hans, Magnetic particles with biologically active receptors, US 6,372,517 B1,16.04.2002
[41] Mosbach, Klaus, Preparation and use of magnetically susceptible polymer particles, US 6,316,235 B1,13.11.2001
[42] Müller-Schulte, Detlef, Preparation of polyvinyl alcoholbased magnetic particles for binding biomolecules, US 6,204,033 B1,20.03.2001
[43] Owen, Charles S., Magnetic-polymer particles, US 5,866,099,02.02.1999
[44] Rauch, Hubert, Method for spray drying polymer emulsions and solutions, US 4,892,932,09.01.1990
[45] Sucholeiki, Irving, Composite magnetic beads, US 5,834,121 A,10.11.1998
[46] The Resinous Product&Chemical Company, Synthetic resinous materials and process of making same, US 595,366,20.11.1943
[47] Tricot, Marc, Process for the preparation of magnetic beads of vinylaromatic polymers, US 4,339,337,13.07.1982
[48] Ugelstad, John, Process for preparing magnetic polymer particles, US 4,774,265,27.09.1988
[49] Ugelstad, John, Process for the production of magnetic polymer particles, EP 0 106 873 B1,29.01.1986
[50] Wang, Chao-Huei J., Magnetically responsive fluorescent polymer particles, US 5,395,688 A,07.03.1995
[51] Whitehead, Roy A., Magnetic particles for use in separations, US 4,554,088,19.11.1985
[52] Yamashita, Hiroshi,Surface treatment method for magnetic particles, US 4,620,987,04.11.1986

3e Discussion Session

Roshan Jachuck

Path forward for accelerating deployment of novel separation processes

4a Microreactor Keynote Address

Asterios Gavriilidis

The state of the art and challenges facing microreaction technology will be discussed

HIGH THROUGHPUT SCREENING OF LOW TEMPERATURE CO OXIDATION AND VOC COMBUSTION CATALYSTS

Stephen Cypes, H. Sam Bergh, Alfred Hagemeyer, Zachary Hogan, Andreas Lesik, Valery Sokolovskii, Guido Streukens, W. Henry Weinberg

Symyx Technologies, Inc., 3100 Central Expressway, Santa Clara, CA 95051, USA, www.symyx.com, scypes@symyx.com

Introduction

The catalytic oxidation of carbon monoxide to carbon dioxide is a key process for respiratory protection, industrial air purification, automotive emissions control and CO clean up in flue gases and fuel cells. Specifically for automotive emissions control, CO catalytic oxidation activity along with VOC combustion and DeNOx catalytic processes are important for cold-start emissions control using 3-way catalytic converters. Extensive research has focused on the improvement of catalytic activity at low temperatures for these 3 co-current catalytic processes. Numerous catalytic systems have been proposed including Pt, Pd, Rh, Ru, Au, Ag, and Cu, supported on refractory or reducible carriers or dispersed in perovskites. Well known commercial catalyst formulations for room temperature CO oxidation are based on CuMn2O4 (hopcalite) and CuCoAgMnOx mixed oxides however these systems have very low activity for VOC combustion and DeNOx, and are rapidly deactivated in the presence of moisture.

High-throughput and combinatorial approaches have been applied to the discovery of more efficient CO oxidation catalysts for various applications. The Symyx screening approach for these reactions was based on a hierarchy of qualitative or semi-quantitative primary screens for discovery of hits and quantitative secondary screens for confirmation and scale up of leads. For CO oxidation, primary screening was carried out by parallel IR thermography capable of screening >10,000 catalysts per month at microgram quantities by careful thermal design of the reactor system. Multi-channel fixed bed reactors equipped with imaging reflection FTIR spectroscopy or GC were used for secondary screening. The multi-channel fixed bed reactors were also used to screen the confirmed CO oxidation hits for VOC combustion catalytic activity. A novel noble-metal free ternary CO oxidation catalytic system, initially discovered using the IR thermography primary screen, will be shown to be an effective catalyst for room-temperature CO oxidation. In addition, optimization of this system shows lower combined CO oxidation and VOC combustion lightoff temperatures as compared to a "standard" Pt system, perhaps indicating the potential for this novel system in 3-way catalytic converters.

Experimental

Our integrated workflow for high-throughput combinatorial synthesis and screening has been described extensively elsewhere [1-2], and the specific workflow employed in this discovery program is shown schematically in Figure 1. This workflow allows for quick identification of active and selective catalyst lead compounds within a large compositional space for low temperature CO oxidation and VOC combustion.

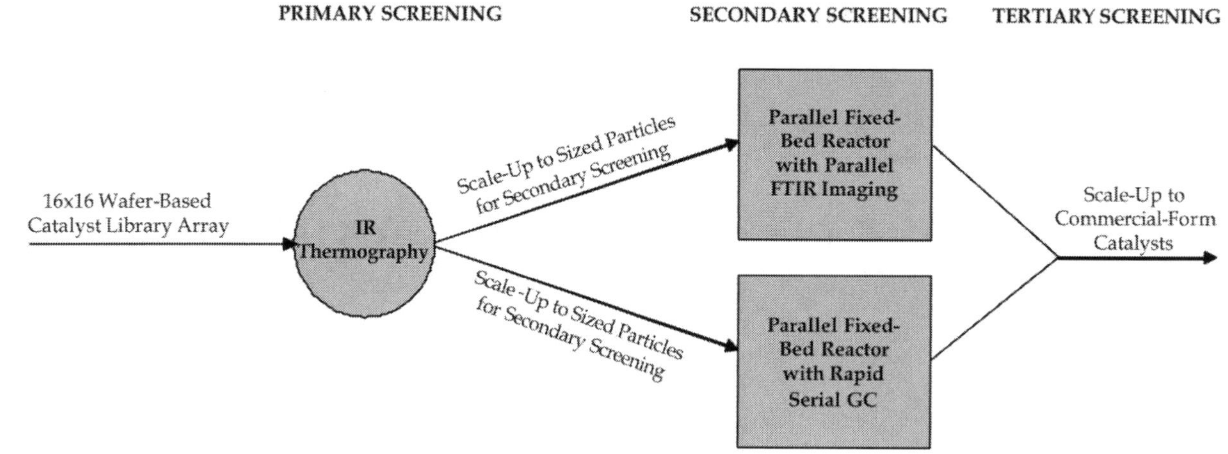

Figure 1. *Schematic of an example Symyx workflow for heterogeneous catalysis discovery used for CO oxidation and VOC combustion catalyst discovery.*

Wafer Synthesis

Wafer-based catalyst libraries were designed using Symyx' Library Studio® software, and synthesized using Symyx' Impressionist® software [3]. Commercial 4" round quartz wafers were cleaned, silanized, and bead blasted through masks to produce 16x16 arrays of wells for catalyst synthesis. These wells were then optionally pre-coated with a metal oxide carrier or commercial support layer using a slurry dispensing technique. Representations of catalyst "library substrates" (i.e. wafers) were defined within Library Studio® as arrays of wells to which a portion of premixed metal precursor solutions are transferred. Once the library design was complete, Library Studio® generated a text file containing instructions that Impressionist® software used to control liquid dispensing robots. These robots physically prepared the wafer libraries. External standards (0.5% Pt on alumina) were slurry-dispensed in several first/last row and last column wells. The remaining spots in first row/last column were left blank to correct for the background signal during data processing. Wells with sample contained ~250 μg catalyst each, resulting in a catalyst film of ~10 micron thickness and 4mm in diameter.

IR Thermography (Primary Screening)

The technique of IR thermography has previously been applied to the discovery and optimization of heterogeneous catalysts [4-8], specifically for CO oxidation as well as total combustion of hydrocarbons, and means of emissivity corrections have been proposed.

The IR thermography system used in these reactions consists of a reactor chamber, a downward-looking IR imaging camera (with associated optics and readout electronics), as well as a gas feed manifold and a means to heat the catalysts to be screened within the reactor, shown schematically in Figure 2. The current

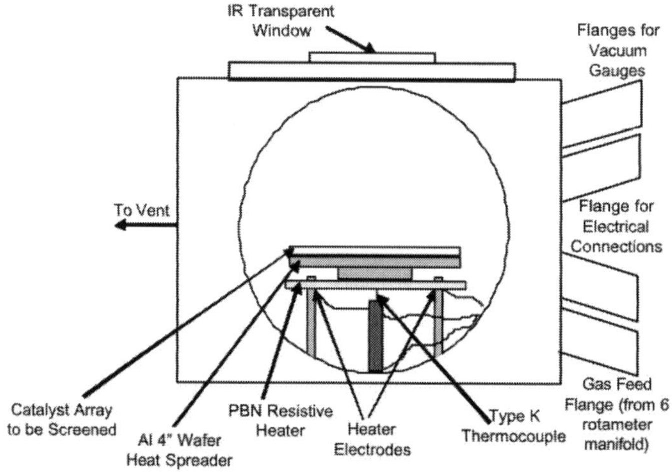

Figure 2. *Schematic of reactor chamber at Symyx for primary screening using IR thermography.*

reactor and optical system can screen reactions under relevant reactions conditions. IR detection from the catalyst array is achieved by a 256 x 256 pixel InSb detector with variable electronic shutter speeds and integration times.

To screen a catalyst library for CO oxidation activity, a candidate wafer is placed on the wafer heater in the reactor chamber and heated to 105°C with 1 SLM clean, dry air (CDA) flow into the reactor chamber. A thermographic image of the candidate wafer is taken by averaging 20 frames. The wafer is then heated to 115°C with 1 SLM CDA flow and another 20-frame average image is taken. A linear calibration of temperature versus pixel response is assumed for each pixel in the detector, and this two-point calibration acts as the non-uniformity correction to account for varying emissivities of materials on the wafer. The wafer is then cooled to 110°C, the desired reaction temperature, and 2% CO in CDA is fed to the reactor chamber at 1 SLM. The candidate catalysts are allowed to equilibrate in the chamber for ~30 minutes, and a 20-frame average image is taken of the catalysts in the presence of CO. The reactor is then purged by feeding 1 SLM of CDA for ~30 minutes while keeping the wafer at 110°C, and a final 20-frame average image is taken. This final image is used to ensure all "hot" catalysts return to the wafer setpoint temperature upon evacuating CO from the chamber, a sign that the presence of CO caused a true temperature rise on the catalyst rather than an irreversible emissivity change of the material.

Parallel Fixed-Bed Reactor with FTIR Imaging or GC (Secondary Screening)

Two separate secondary screening reactors were employed in order to validate and quantify the hits identified from IR thermography. The first reactor consisted of 6 parallel fixed-beds with a flow-distributed feed, in which each of the 6 reactor effluent gases flowed to hollow flexible IR waveguides for FTIR imaging analysis [9]. 2% CO in CDA acted as the feed to each reactor at a space velocity of 50,000 1/hr. The reactor temperature was increased from 40°C to 200°C in 10-20°C increments in order to produce light-off curves for CO oxidation. Since FTIR imaging allows for the simultaneous analysis of CO to CO_2 conversion from all 6 reactors, light-off curves could be produced quickly (~2 hours) for a set of 6 catalyst candidates.

The second reactor consisted of 8 parallel fixed-beds with a flow-distributed feed, in which a stream selection valve selected one of the reactor effluents for rapid serial GC analysis [10]. The feed consisted of 1% CO and/or 0.15% propylene in CDA at a space velocity of 25,000 1/hr. The reactor temperature was increased from 40°C to 200°C in 10-20°C increments in order to produce light-off curves for CO oxidation and propylene combustion both as pure components in CDA as well as a mixed CO-propylene-CDA feed. GC analysis used a TCD for quantitative detection of CO, CO_2, H_2O, propylene and O_2. Light off curves could be produced for the CO oxidation and VOC combustion analysis of all 8 candidate catalysts in ~12 hours, via automated Symyx' Impressionist® software control.

Results and Discussion

The IR thermography reactor was initially used to screen pure and binary metals on supports and bulk oxides, as a function of variables such as precursor salt and calcinations/reduction conditions. Figure 3 shows a typical thermographic image of a 256-element library of candidate catalysts once CO is introduced into the reactor chamber. Clearly, "hits" are identified for metal "A" (all precursors) on bulk oxide support "B", with total temperature rise comparable to the Pt standards.

Additional libraries were synthesized and screened via IR thermography for potential ternary metal systems (Figure 4), as well as for the optimization of the support (both material as well as surface area), precursors, and calcinations/reduction conditions. Greater than 20,000 catalysts within several months were screened for CO oxidation activity via this primary screen.

The best "hits" from the IR thermography primary screen were initially confirmed to be active using the 6-channel parallel fixed-bed reactor coupled to FTIR imaging. Figure 5 shows the results of a typical experiment from this secondary screen. The results are shown in two formats: light-off curves for the six catalysts, as well as pairs of images representing the total absorbance peak area for either CO or CO_2 from each reactor at several reactor temperature setpoints. The images result in a quick visual determination of which reactor effluent reaches light-off of CO fastest, and hence is the most active catalyst.

Additional secondary screening of catalyst "hits," as well as optimization of the catalyst for VOC combustion, was performed in the 8-channel parallel fixed-bed reactor with GC analytics. Figure 6 shows the resulting light-off curves for both propylene and CO for several potential catalysts using a mixed propylene-CO-CDA feed. It can be seen from these plots that 6 out of the 7

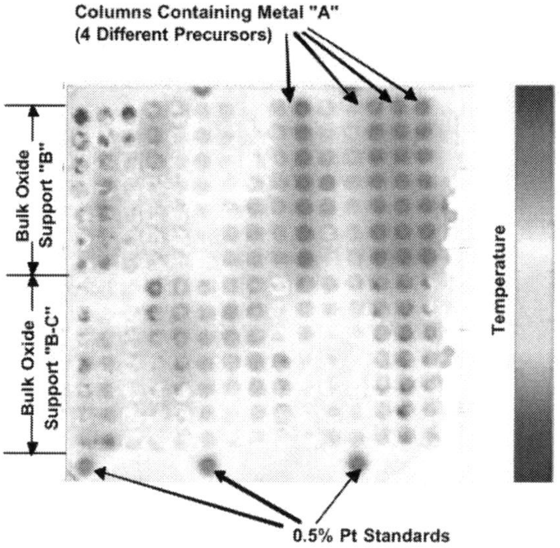

Figure 3. *Early "hit" identification for Metal "A" on bulk oxide "B" using IR thermography primary screen.*

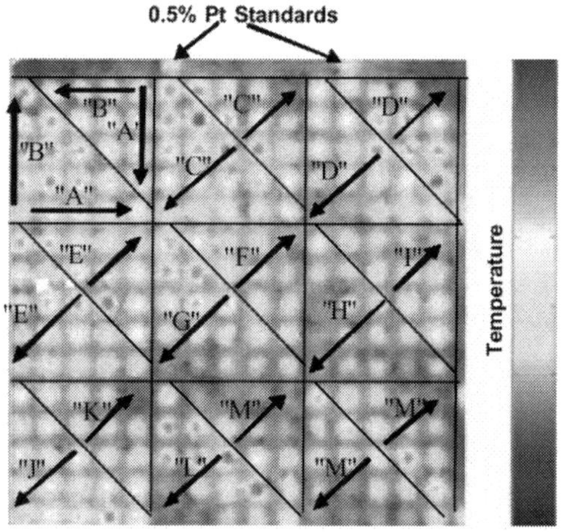

Figure 4. *Optimized libraries for quick identification of potential ternary metal systems.*

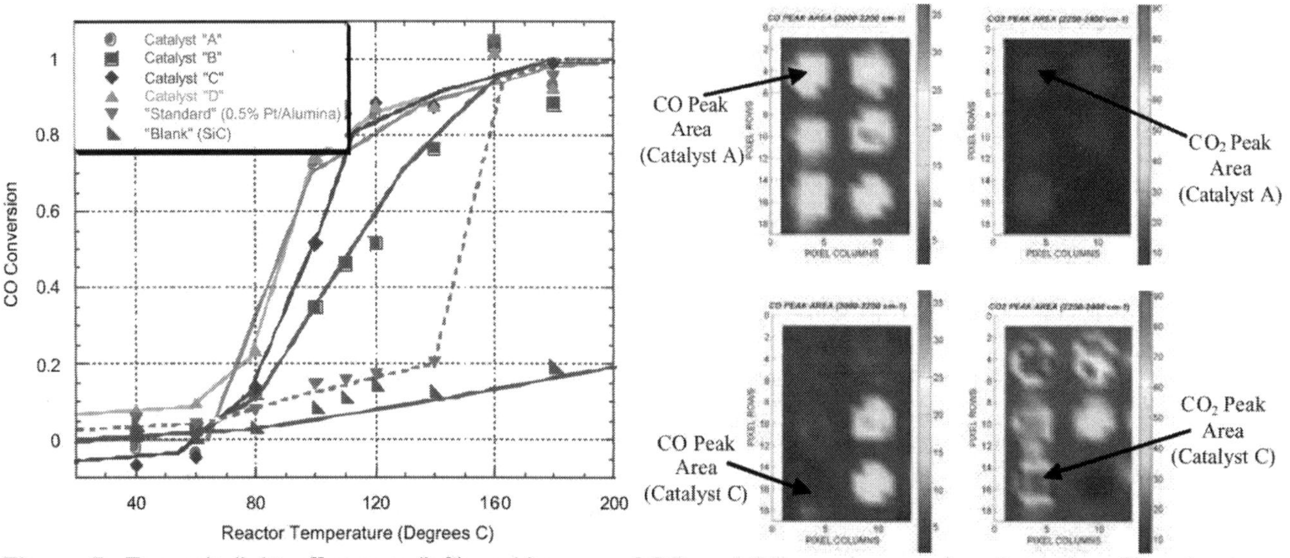

Figure 5. *Example light-off curves (left) and images of CO and CO_2 peak areas for all reactor effluents at 40°C (top right) and 140°C (bottom right).*

noble-metal free experimental catalysts reveal a lower light-off temperature for both CO and propylene as compared to 0.5% Pt on alumina.

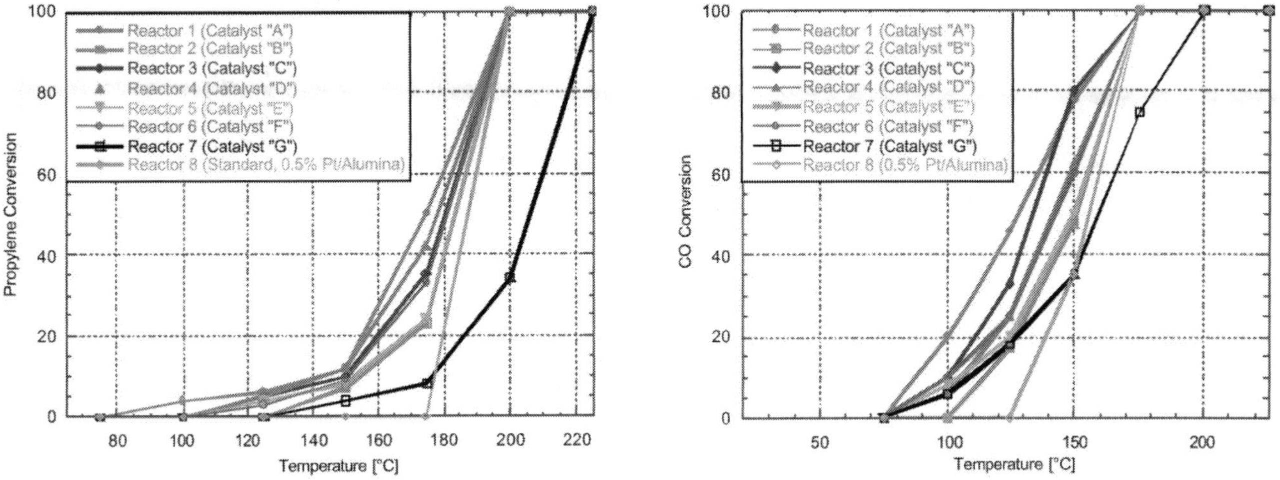

Figure 6. *Light-off curves for propylene (left) and CO (right) using 8-channel parallel fixed-bed reactor with mixed CO-propylene-CDA feed. 6 of the 7 noble-metal free catalysts result in lower light-off than standard Pt catalvst.*

Conclusions

The Symyx workflow presented here was capable of identifying 4 completely independent novel groups of noble-metal free CO oxidation and VOC combustion catalysts. These discoveries were accomplished by screening >20,000 potential catalysts via a primary screen utilizing IR thermography, and further optimizing "hits" via secondary screens utilizing parallel fixed-bed reactors with either FTIR imaging or GC analytics. Additional high-throughput screening and optimization for DeNOx activity could lead to novel classes of catalysts suitable for 3-way catalytic converters. The discoveries made in this study are suitable for further development towards low-temperature CO oxidation and total VOC combustion.

References

[1] A. Hagemeyer, B. Jandeleit, Y. Liu, D.M. Poojary, H.W. Turner, A.F. Volpe Jr, W.H. Weinberg, "Applications of combinatorial methods in catalysis", Appl. Catal. A 221 **(2001)** 23-43.

[2] H. Bergh, in "High Throughput Screening in Chemical Catalysis: Technologies, Strategies and Applications" (eds. A. Hagemeyer, P. Strasser, A. Volpe), Wiley-VCH, Weinheim, **2004**.

[3] US Patent No.'s: 6004617, 5985356, 6326090, 6627571. Additional US and foreign patents issued and pending.

[4] A. Holzwarth, H. Schmidt, W. Maier, "Detection of Catalytic Activity in Combinatorial Libraries of Heterogeneous Catalysts by IR Thermography." *Angew. Chem. Int. Ed.*, 37(19) **1998** 2644.

[5] W. Li, F.J. Gracia, E.E. Wolf, *Catalysis Today*, 81 **(2003)** 437-447.

[6] G. Kirsten , W.F. Maier, in "High Throughput Screening in Chemical Catalysis: Technologies, Strategies and Applications" (eds. A. Hagemeyer, P. Strasser, A. Volpe), Wiley-VCH, Weinheim, **2004**.

[7] U.S. Patent No.'s: 6410331, 6514764, 6063633, 6333196, 6841127. Additional US and foreign patents issued and pending.

[8] Willson, R. C. World Patent PCT WO 97/32208, 1997

[9] U.S. Patent No.'s: 6410331, 6514764, 6149882, 6623968, 6623970. Additional US and foreign patents issued and pending.

[10] U.S. Patent No.'s: 6410331, 6514764, 6149882, 6623968, 6623967. Additional US and foreign patents issued and pending.

4c Production of Hydrogen Peroxide by Controlled H2/O2 Reaction in a Microchannel Reactor

Adeniyi Lawal, Emmanuel Dada, Raghunath Halder, Sunitha Tadepalli, Woo Lee, Yury Voloshin

Hydrogen peroxide was formed in a packed-bed microreactor in a three-phase process involving the direct combination of hydrogen and oxygen over a solid catalyst with the hydrogen peroxide dissolved in a liquid solvent. Hydrogen and air were combined in all proportions including explosive regime over Pd catalyst supported on various oxides. The goal of the work presented here is to elucidate the mass-transfer mechanism by identifying the different controlling regimes of this reaction as a prelude to the determination of the intrinsic kinetics in the packed-bed microreactor. Absence of internal mass transfer limitation was confirmed by measuring the reaction rates with catalyst particles of different sizes. Two approaches were adopted for the determination of the influence of external mass transfer resistance on the kinetics of the reaction. In the first approach, the flow velocity was varied while keeping the residence time constant. The effect of velocity on reaction rates was an indication of external mass transfer limitation. The dominant controlling regime was confirmed by comparing the observed reaction rates with calculated external mass transfer rates using estimated mass transfer coefficients from the literature. In the second approach, the effect of temperature on the reaction rates over a range of temperatures (Arrhenius curve) was determined, keeping all the other processing conditions fixed. Comparison of reaction rates with different residence times but at constant flow velocity served as another confirmation of the controlling regime.

The use of solid-supported reagents within microfabricated devices

Charlotte Wiles

Department of Chemistry, The University of Hull, Cottingham Road, Hull, HU6 7 RX, UK.

Abstract

As a result of increasing environmental pressure, the chemical industry as a whole are exploring many routes in order to improve both the cleanliness and efficiency of many synthetic processes. One such approach is the application of micro reaction technology, which enables reactions to be performed more rapidly, efficiently and selectively than traditional batch-scale reactions. Although many groups have demonstrated the advantages of performing organic reactions within micro fabricated devices, few have addressed the problems associated with the purification of reaction products prepared using continuously flowing systems. In order to tackle this problem, we investigated the incorporation of an array of solid-supported reagents into a miniaturised flow reactor.[1] Using this methodology, we demonstrate the synthesis of dimethyl acetals, aldehydes, carboxylic acids and α,β-unsaturated ketones in high purity without the need for additional product purification.

Introduction

Compared to solid-phase techniques, where reaction products cannot be fully characterised until they are cleaved from the support, the use of solid-supported reagents is advantageous as starting materials, intermediates and products remain in solution, thus enabling the reaction to be monitored with time (Figure 1). In addition, as the supported reagent can be easily removed from the reaction mixture, excess amounts can be employed in order to drive the reaction to completion. Although solid-supported reagents have many advantages over their solution phase counterparts, one limitation is support degradation (as a result of stirring or shaking the reaction mixture) which makes reagent recycling difficult. Therefore by performing reactions in continuous flow reactors, such as the one described herein, the support undergoes minimal degradation, which results in extended reagent lifetimes and system reproducibility.[2] Automation of the technique therefore enables the high-throughput synthesis of analytically pure compounds, suitable for the fine chemical industry or combinatorial applications. With these factors in mind, we propose that by incorporating a series of solid-supported catalysts and regents into miniaturised flow reactors, problems such as corrosion of reactor vessels, generation of acidic waste and the inability to efficiently recover and recycle the catalyst can be addressed. In order to demonstrate the advantages associated with the proposed technique, we demonstrate the acid catalysed synthesis of dimethyl acetals, base catalysed Knoevenagel condensation and the stoichiometric oxidation of primary alcohols.

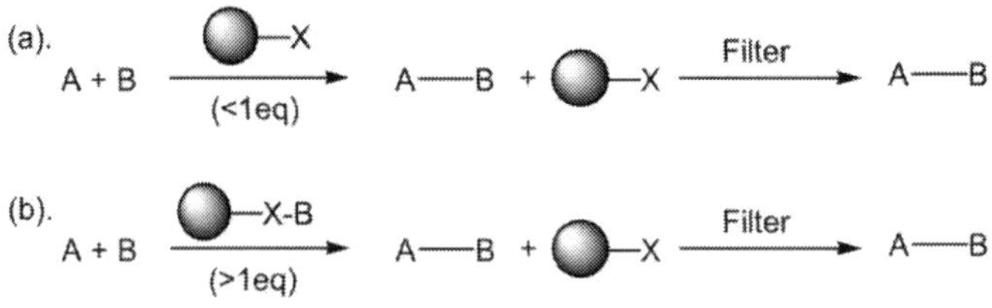

Figure 1. Comparison of (a). catalytic and (b). stoichiometric solid-supported reagents.

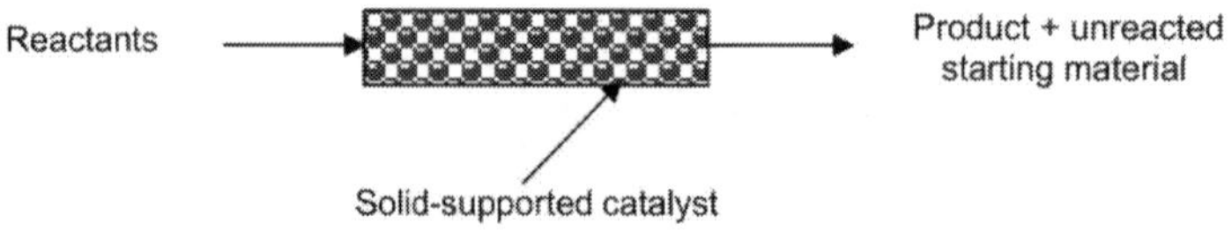

Figure 2. Schematic illustrating the use of solid-supported catalysts in a continuous flow reactor.

To conduct a reaction, the starting materials are passed through a solid-supported reagent or catalyst and the reaction products collected at the outlet (Figure 2). The reaction mixture is then analysed by GC-MS whereby conversion of starting material to product is determined. If any residual starting material is observed the reaction is repeated, this time passing the reagents over the support at a slower flow rate, thus having the effect of increasing the residence time within the reactor. When successfully optimised, the devices are operated continuously in order to prepare a sufficient quantity of product for analysis by NMR spectroscopy and if required, elemental analysis. Using this approach, work-up is extremely simple, consisting of concentrating *in vacuo* followed by chromatographic analysis. By optimising the flow rate and hence residence time within the reactor, it is possible to obtain complete conversion of starting material to product in a single pass through the device.

Experimental

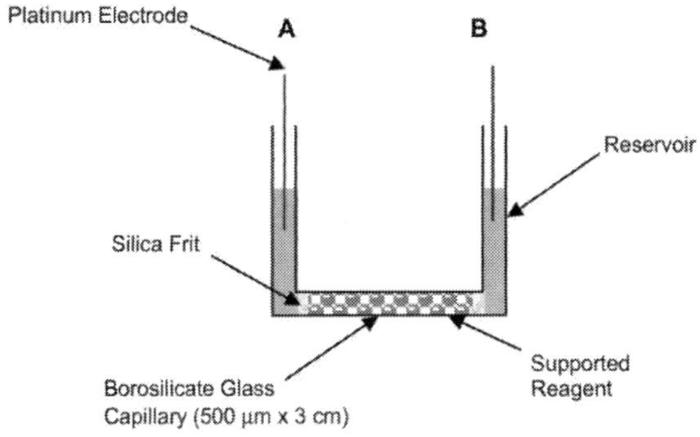

Figure 3. Reaction set-up for the evaluation of supported reagents.

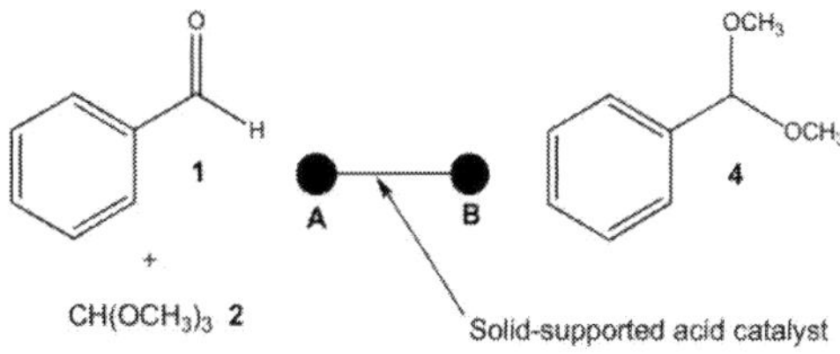

Figure 4. Synthesis of dimethyl acetal **4** in an EOF-based micro reactor.

In order to evaluate the use of solid-supported reagents within an EOF-based system, a miniaturised flow reactor was designed (Figure 3); this approach is advantageous as it not only enables reagents to be packed with ease, but also provided a relatively inexpensive, versatile system. For initial investigations, electroosmotic flow (EOF)[3] was selected as the pumping mechanism as it is easy to use, requires no mechanical parts, provides reproducible pulse-free flow and most importantly with respect to packed systems, generates minimal back-pressure.[4] As Figure 3 illustrates, micro porous silica frits are placed at both ends of the borosilicate glass capillary in order to retain the supported reagent. The capillary is then primed with the desired solvent system and attached to the two glass reservoirs. The reagents can then be manipulated through the device by application of a voltage to the platinum electrodes placed in the reagent reservoirs.

Results and Discussion

Polymer-supported acid catalysis

Acetals are one of the most common carbonyl protecting groups, prepared by the treatment of aldehydes (or ketones) with alcohols (or orthoformates) in the presence of an acid catalyst (Scheme 1). Although triflic acid and *p*-toluenesulfonic acid are generally used, other catalysts include ferric chloride, ammonium nitrate and ethanolic hydrogen chloride.[5] In addition, numerous examples of solid-supported acid catalysts have been applied to the synthesis of acetals, these include Amberlite resin, polymer-supported lanthanides and Nafion-H.[6] With this in mind, we investigated the synthesis and deprotection of an array of dimethyl acetals within an EOF-based flow reactor. By flowing a premixed solution of benzaldehyde **1** and trimethylorthoformate **2** in anhydrous MeCN (1.0 M) over a packed bed of Amberlyst-15 **3** at a flow rate of 1.75 µl min^{-1} (300 V cm^{-1}), resulted in 100 % conversion of benzaldehyde **1** to dimethoxymethyl benzene **4** (Figure 4).

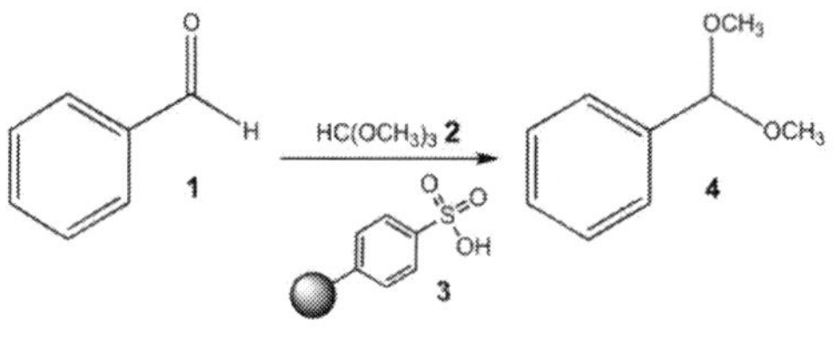

Scheme 1. General scheme illustrating the acid catalysed synthesis of dimethoxymethyl benzene **4**.

In order to demonstrate both system reproducibility and the continuous synthesis of dimethyl acetal **4** the reaction was repeated a further 14 times (2.5 hr), whereby conversions of > 99.8 % (% RSD = 0.13) were obtained. After analysis by GC-MS, the reaction products were collected and concentrated *in vacuo*, to afford dimethoxymethyl benzene **4** as a pale yellow oil (0.025 g, 96.6 %); subsequent analysis of the crude product by NMR spectroscopy confirmed product purity. In order to demonstrate the generality of the technique, the reaction was subsequently repeated synthesising dimethyl acetals **5** to **14** (Table 1) in > 99.6 % conversion. Having demonstrated the successful incorporation of Amberlyst-15 **3** into an EOF-based flow reactor, the investigation was extended to the use of ytterbium (III) polystyrylsulfonate **15** and polymer-supported *p*-toluenesufonic acid **16** whereby an array of acetals were successfully synthesised in > 94.7 % yield (Tables 2a and 2b).

Table 1. Summary of the conversions obtained for the synthesis of dimethyl acetals **4** to **14**.

Product	Flow Rate ($\mu l\ min^{-1}$)	Conversion[a] (%)	RSD (%)	Yield (%)	Product	Flow Rate ($\mu l\ min^{-1}$)	Conversion[a] (%)	RSD (%)	Yield (%)
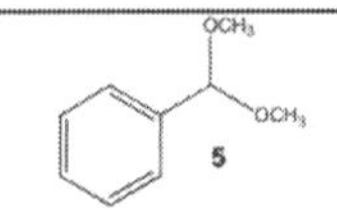	1.75	99.77	0.13	96.6	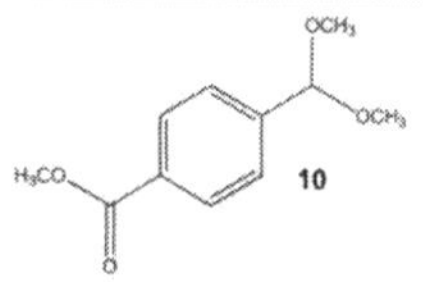	0.60	99.86	0.08	95.3
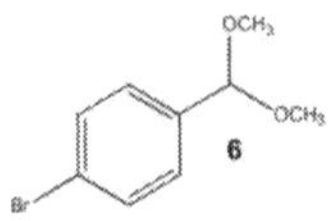	1.00	99.92	0.22	96.8	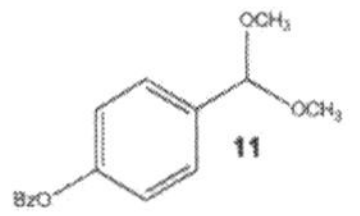	0.35	99.70	0.15	98.13
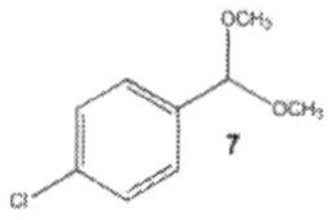	1.60	99.78	0.15	98.0	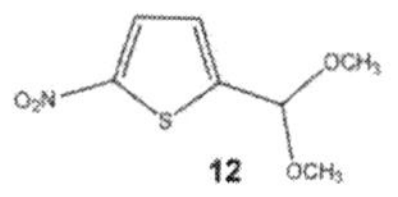	2.00	99.88	0.93	97.5
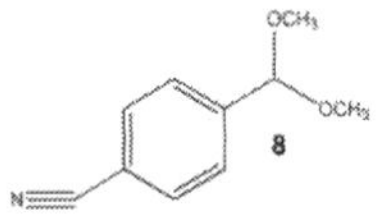	2.00	99.64	0.90	97.5	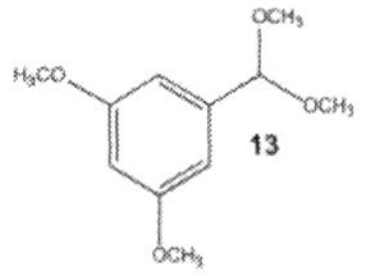	0.50	99.84	0.24	98.4
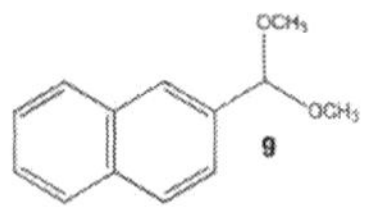	1.40	99.83	0.26	95.2	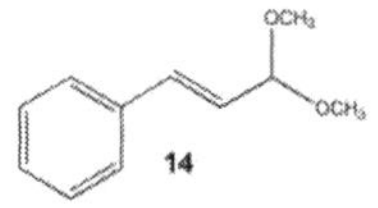	1.30	99.65	0.29	95.4

[a] ≥ 10 replicates were performed for each compound.

Along with the ability to introduce a protecting group, it is important to be able to cleanly, selectively and efficiently remove it; with this in mind, we investigated the acid catalysed deacetalisation of dimethyl acetals.[7] By operating the device at flow rates < 1.00 $\mu l\ min^{-1}$, acetals **4** to **14** were successfully deprotected to afford the respective aldehyde in > 99.8 % conversion (> 94.8 % yield).

Table 2. Conversions obtained using (a). ytterbium (III) polystyryl sulfonate **15** and polymer-supported *p*-toluene sulfonic acid **16**.

(a).

Product	Flow Rate (μl min^{-1})	Conversion[a] (%)	RSD (%)	Yield (%)
5	0.40	99.72	0.13	94.7
6	0.40	99.96	0.06	98.8
7	0.28	99.97	0.08	96.3
8	0.52	99.92	0.05	96.8
9	0.40	99.87	0.15	97.7
10	0.40	99.72	0.06	97.2
11	0.70	99.88	0.03	98.7
12	0.55	99.83	0.08	95.5
13	0.95	99.83	0.12	98.6
14	0.90	99.64	0.14	96.1

(b).

Product	Flow Rate (μl min^{-1})	Conversion[a] (%)	RSD (%)	Yield (%)
5	1.10	99.80	0.20	96.8
6	0.30	99.86	0.27	96.0
7	1.40	99.85	0.19	97.6
8	0.79	99.93	0.21	95.9
9	1.00	99.77	0.21	98.3
10	0.70	99.74	0.12	95.8
11	0.60	99.64	0.25	95.7
12	5.00	99.87	0.17	97.8
13	1.00	99.85	0.15	94.9
14	1.70	99.70	0.11	98.5

[a] ≥ 15 replicates were performed for each compound.

Having demonstrated the successful incorporation of supported acids into an EOF-based continuous flow reactor, we subsequently investigated a series of supported bases using the Knoevenagel condensation as a model reaction.

Silica-supported base catalysis

As Scheme 2 illustrates, the Knoevenagel condensation is the base catalysed condensation of an aldehyde (or ketone) with compounds that posses an activated methylene group, *i.e.* cyano, acyl or nitro, to afford an unsaturated product. The main disadvantage associated with this reaction is that even if the reaction proceeds to completion, the unsaturated products require purification in order to remove any residual organic base and its respective salt. Consequently, by incorporating a supported base into a continuous flow reactor we proposed that analytically pure α,β-unsaturated compounds could be synthesised without the need for additional purification. In order to perform a reaction, a solution of benzaldehyde **1** and ethylcyanoacetate **17** (1.0 M in anhydrous MeCN) was passed over 3-(1-piperazino) propyl-functionalised silica gel **18** at a flow rate of 0.5 μl min^{-1} (333 V cm^{-1}), resulting in 98.3 % conversion to 2-cyano-3-phenyl acrylic acid ethyl ester **19**. Again, in order to demonstrate the use of this approach for continuous synthesis, the reactor was operated continually over a period of 4.75 hr (14 runs), whereby 0.025 g (0.124 mmol, 98.9 %) of product **19** was obtained (Table 3 and Figure 5). The reaction was subsequently repeated for 4-bromobenzaldehyde **20**, 3,5-dimethoxybenzaldehyde **21** and 4-benzyloxybenzaldehyde **22** to afford the respective unsaturated ketones **23**, **24** and **25** in > 96.3 % (Table 4).

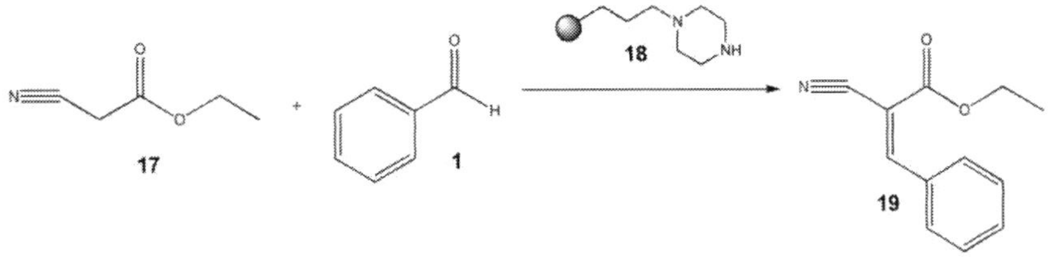

Scheme 2. Illustration of the reaction of an activated methylene with aldehydes in the presence of a supported base **18**.

Having successfully demonstrated the preparation of an array of condensation products, the technique was extended to the synthesis of 2-benzylidene malononitrile **26** (96.9 %), 2-(4-bromobenzylidene)-malononitrile **27** (96.9 %), 2-(3,5-dimethoxybenzylidene)-malononitrile **28** (96.3 %) and 2-(4-benzyloxybenzylidene)-malononitrile **29** (97.3 %), in all cases no by-products were formed.

Table 3. Illustration of device reproducibility for the synthesis of product **19**.

Run No.	Conversion (%)
1	98.3
2	98.5
3	98.3
4	98.3
5	98.4
6	99.2
7	99.1
8	99.1
9	100.0
10	99.6
11	99.3
12	100.0
13	100.0
14	99.2
Mean = 99.1 %, % RSD = 0.65	

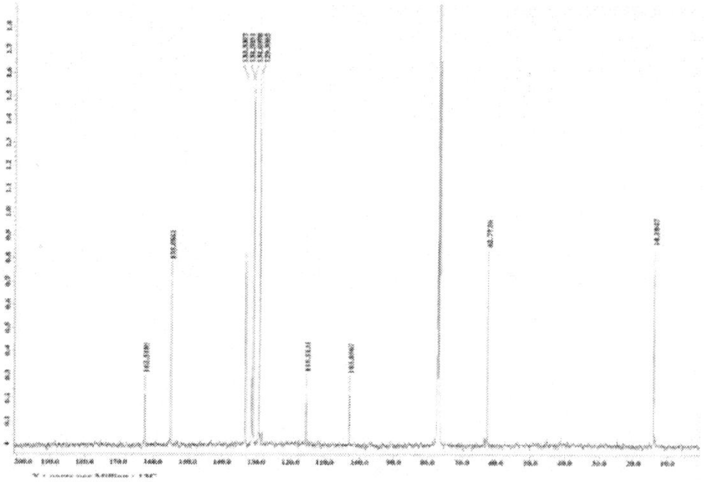

Figure 5. ^{13}C NMR of 2-cyano-3-phenyl acid ethyl ester **19** synthesised in a micro fabricated device.

Table 4. Summary of the conversions obtained in a miniaturised continuous flow reactor using 3-(1-piperazino)propyl-functionalised silica gel **18**.

Product	Applied Field (V cm^{-1})	Flow Rate (μl min^{-1})	Conversion[a] (%)	Product	Applied Field (V cm^{-1})	Flow Rate (μl min^{-1})	Conversion[a] (%)
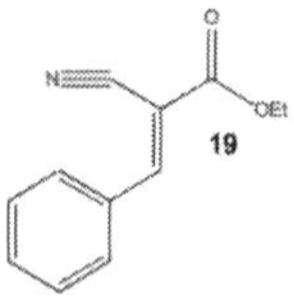 19	333	0.5	99.10	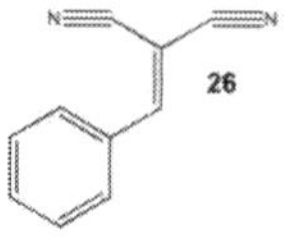 26	167	1.0	96.90
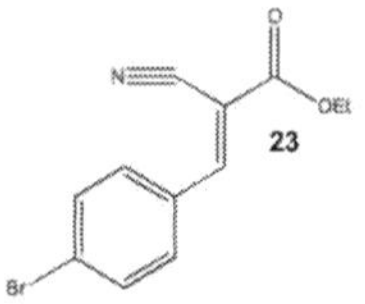 23	333	0.3	99.50	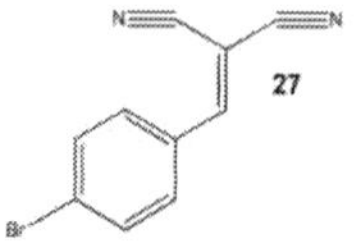 27	167	0.5	96.30
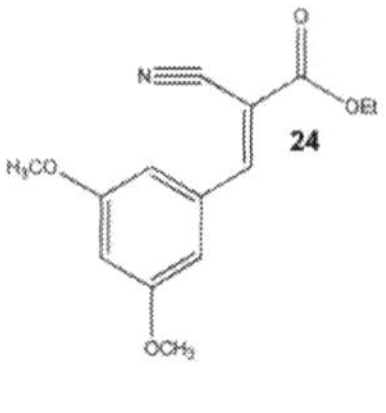 24	333	0.3	94.70	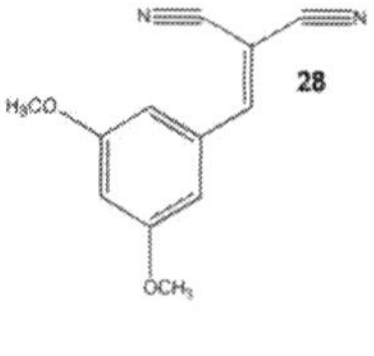 28	167	0.7	97.80
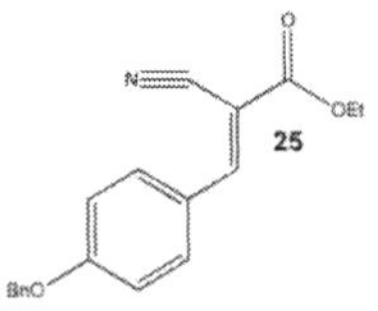 25	333	0.5	95.10	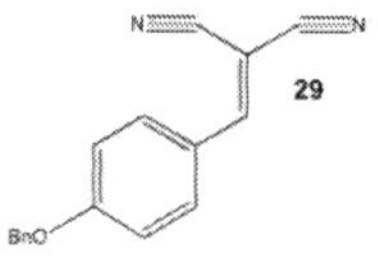 29	167	1.0	99.70

[a] ≥ 10 replicates were performed for each compound.

The technique was subsequently repeated using 3-(dimethylamino)propyl-functionalised silica gel, 3-aminopropyl-functionalised silica gel and 3-(1,3,4,6,7,8-hexahydro-2*H*-pyrimido[1,2-a]-pyrimidino)propyl-functionalised silica gel, whereby 99.4, 100 and 99.3 % conversion to 2-cyano-3-phenyl acrylic acid ethyl ester **19** were obtained respectively (Table 4).

Silica-supported oxidant

To further demonstrate the advantages associated with performing continuous flow synthesis, we extended the investigation to the selective oxidation of a series of primary alcohols. As Scheme 3 illustrates, when a primary alcohol **30** is oxidised precautions must be taken to ensure that the desired aldehyde **1** is not further oxidised to the corresponding carboxylic acid **31**. In practise, this can be achieved by distillation of the aldehyde as it forms; we however propose that by conducting the reaction in a continuous flow reactor, selective synthesis of the aldehyde can be achieved.

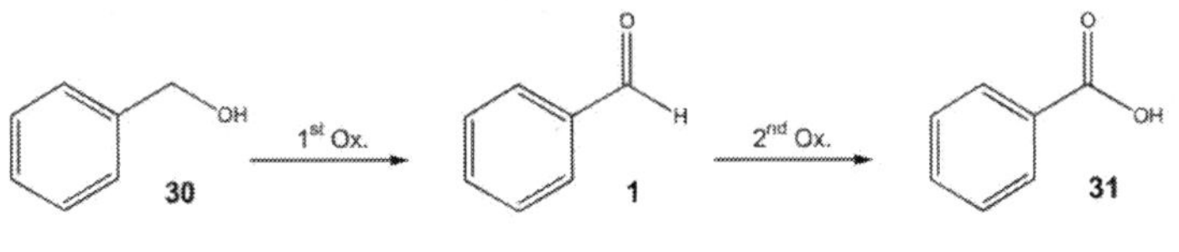

Scheme 3. Scheme illustrating the over oxidation of benzyl alcohol **30** to benzoic acid **31**.

Table 5. Effect of solvent system on Jones reagent **32**.

Solvent system	Chromium release
Water	✓
Acetonitrile	✓
Tetrahydrofuran	✓
Diethyl ether	✓
Dichloromethane	×

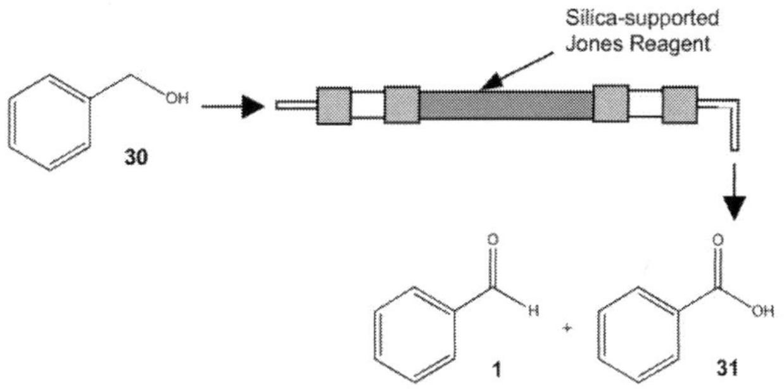

Figure 6. Flow system used for the oxidation of 1^{0}alcohols.

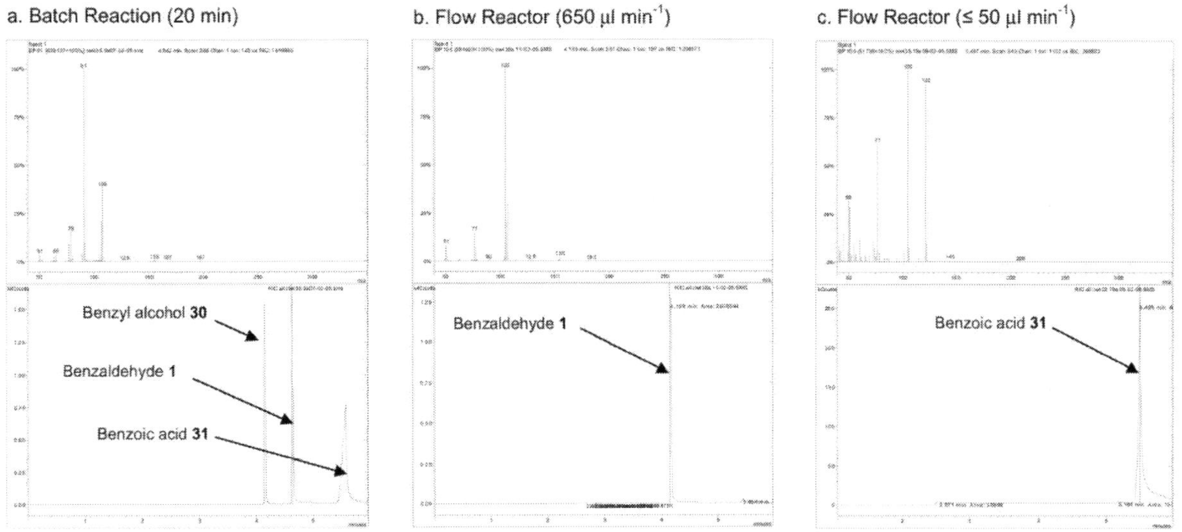

Figure 7. Gas chromatograms illustrating the product selectivity obtained in a flow reactor compared to batch.

To demonstrate the selective oxidation of primary alcohols in a flow reactor, we investigated the use of silica-supported Jones reagent **32**;[8] however, unlike previous examples where an EOF-based device was employed, due to the solvent incompatibilities (Table 5) a pressure-driven system was investigated (Figure 6). Prior to performing a flow reaction, the silica-supported Jones reagent **32** was investigated in batch; as Figure 7 illustrates after stirring for 20 min over-oxidation to benzoic acid **31** was observed. In order to perform a flow reaction, a solution of benzyl alcohol **30** in DCM (0.01 M) was pumped through the device at a flow rate of 650 µl min^{-1} and the reaction products analysed by GC-MS. As Figure 7 illustrates, when operating the device at high flow rates, over-oxidation is prevented and complete conversion to the aldehyde **1** is observed. Interestingly, when the device was operated at 300 μl min^{-1}, the reaction products were a mixture of unreacted starting material **30**, aldehyde **1** and carboxylic acid **31**, however further reducing the flow rate to < 50 μl min^{-1}, resulted in the selective synthesis of the over-oxidation product, benzoic acid **31**.

Having demonstrated the ability to selectively synthesise either benzaldehyde **1** or benzoic acid **31** with no residual benzyl alcohol **30**, the reaction was repeated using 3,5-dimethoxybenzyl alcohol **33**, 4-bromobenzyl alcohol **34**, 4-chlorobenzyl alcohol **35** and 4-cyanobenzyl alcohol **36**. In this case a range of substituted benzyl alcohols were investigated and as Table 6 illustrates excellent selectivity was obtained and no functional group incompatibilities were detected.

Table 6. Table illustrating preliminary results obtained for the oxidation of primary alcohols in a flowing system.

		Conversion (%)	
Alcohol	Flow Rate/μl min^{-1}	Aldehyde	Carboxylic acid
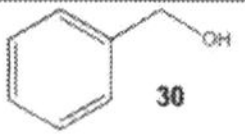 30	650	100	0
	50	0	100
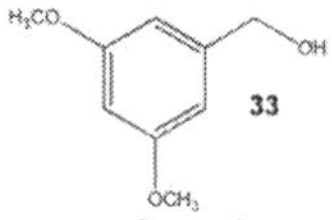 33	650	100	0
	50	0	100
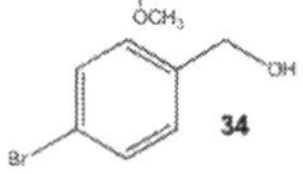 34	650	100	0
	50	0	100
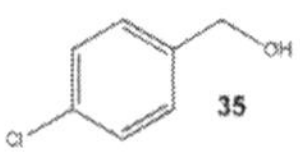 35	650	100	0
	50	0	100
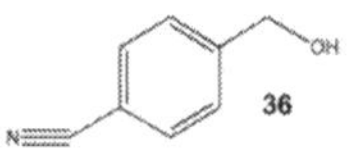 36	650	100	0
	50	0	100

[a] ≥ 10 replicates were performed for each compound.

As a result of the unique reaction conditions obtained in continuous flow systems, we have demonstrated product selectivity not attainable in batch (Figure 7) *i.e.* at fast flow rates aldehydes are synthesised and at slow flow rates carboxylic acids are prepared (Table 6). With this in mind, future work will go on to investigate if the aforementioned reaction selectivity is also attainable in catalytic systems *i.e.* recycling the oxidising agent using a co-oxidant.

Conclusions

Compared to standard batch techniques, the methodology described herein is advantageous as it enables the supported reagents to be recycled without the need for filtration; consequently more consistent results are obtained as no reagent is lost. In addition, the absence of stirring/shaking greatly reduces mechanical degradation of the support, enabling it to be employed for longer and the presence localised concentration gradients enables reactions to be driven to completion without the need to employ large quantities of supported catalyst (2.5 to 5.0 mg). Consequently, reaction conditions can be optimised, enabling small quantities of analytically pure compounds to be prepared in minutes. Alternatively, larger quantities of materials can be synthesised by simply scaling-out the process *i.e.* operating numerous devices. By incorporating a series of supported acids, bases and oxidants, we have demonstrated the synthesis of an array of compounds including acetals, aldehydes, carboxylic acids and α,β-unsaturated compounds. In all cases, conversions of > 95 % were obtained and subsequent analysis of the crude reaction mixtures by NMR spectroscopy confirmed excellent product purity. The ability to prepare pure compounds in sufficient quantities to analyse by NMR spectroscopy is also advantageous as structural information can be obtained without relying on the use of synthetic standards and comparative chromatographic techniques. Applying the methodology described herein, further studies are currently underway within our laboratories to extend both the type of reagent and support employed, enabling more complex syntheses to be evaluated.

Acknowledgements

Full financial support provided by the EPSRC (Grant No. GR/S34106/01) (C.W.) is gratefully acknowledged. Dr. Paul Watts and Prof. Steve Haswell are thanked for their supervision and guidance and Mike Bailey (The University of Hull) assistance with device fabrication is gratefully acknowledged.

References

1. Wiles, C.; Watts, P.; Haswell, S. J. *Tetrahedron* **2004**, *60,* 8421.
2. Hodge, P. *Curr. Opin. Chem. Biol.* **2003**, *7*, 362.
3. Rice, C. L.; Whitehead, R. *J. Phys. Chem.* **1965**, *69*, 4017.
4. Fletcher, P. D. I.; Haswell, S. J.; Paunov, V. N. *Analyst* **1999**,*124*, 1273.
5. Patwardhan, S. A.; Dev, S. *Synthesis* **1974**, 348.
6. Olah, G. A.; Narang, S. C.; Mediar, D.; Salem, G. F. *Synthesis* **1981**, 282.
7. Coppola, G. M. *Synthesis* **1984**, 1021.
8. Ali, M. H.; Wiggin, C. J. *Synth. Commun.* **2001**, *31*, 1389.

4e Olefins by High Intensity Oxidation of Ethane

Terry Mazanec

Velocys, in partnership with Dow Chemical and PNNL, is developing a breakthrough process for the production of ethylene by selectively oxidizing ethane in a microchannel reactor in a project sponsored by the DOE. The process is expected to provide benefits in energy utilization, feedstock utilization, and capital expenditure compared to conventional ethylene technology.

The Velocys® microchannel reactor technology carefully manages heat generation and consumption to give an optimal temperature profile in the oxidation of ethane to ethylene. The oxidation process has the potential to generate higher selectivities, conversions and throughputs compared to conventional steam cracking since the oxidation reaction is not equilibrium limited. Control of kinetics by close thermal control provides an advantage for the Velocys process.

Phase 1 of the project is devoted to the development of the fundamental underpinnings of the technology. Phase 2 will focus on improving reactor design, building on the experience from the single channel tests and modeling results

4f Microchannel Reactor System for Catalytic Hydrogenation of o-nitroanisole to o-anisidine

Adeniyi Lawal, Donald Kientzler, Luke Achenie, Raghunath Halder, Ronald S Besser, Sunitha Tadepalli, Woo Lee

In this work, a three phase solid catalyzed hydrogenation reaction of o-nitroanisole to produce o-anisidine in methanol was studied in a microreactor as a model pharmaceutical reaction. In the pharmaceutical industry, hydrogenation reactions constitute about 15-20% of the reactions and are generally carried out in batch reactors. Ineffective mass transfer and non-uniform temperature distribution in the batch reactor can have deleterious effects on reactor performance including the formation of undesired side products. These side products lead to product impurity necessitating costly separation processes. By using a microreactor for conducting such highly exothermic reactions, significantly improved heat and mass transfer can be achieved because of the high surface-to-volume ratio in these reactors. The product selectivity can also increase due to short reaction time because of the attainment of close to intrinsic kinetics in the microreactor, especially in reactions where products can further react catalytically to form side products.

The main objective of this study is to compare the performance of the microreactor with a batch reactor using this model reaction. Experiments were conducted in a packed bed micro-reactor using Pd catalyst on different supports to determine the best support for the hydrogenation reaction. It was found that Pd supported on zeolite resulted in higher yield and selectivity compared to Pd on other supports. The effects of different processing conditions, e.g., reactor temperature, pressure, nitroanisole concentration in the feed on the conversion, yield and selectivity have been studied using the Pd supported on zeolite catalyst. Similar experiments were conducted in the batch reactor using the same catalyst for the comparison study.

4g Discussion Session

Roshan Jachuck

Facilitated open discussion with microreaction session attendees

5a Business Opportunities for PI

Thomas McHugh

Nexpress Inc. has been involved in adopting the concept of process intensification (PI) for achieving energy efficiency, improvement in product quality and reduction in foot print. This presentation will highlight the business opportunity offered by PI technology whilst addressing the industrial challenges for adopting an emerging technology.

5b Emerging Trends for PI Equipment

Mike Jones

Protensive is one of the world's leading technology provider for PI technologies. It has expertise in equipment, as well as process technology for a wide spectrum of process industry. This presentation will highlight the emerging trends, scope and the perceived way forwards for PI.

5c Discussion Group

Roshan Jachuck

An open, facilitated discussion session on state-of-play, future challenges and needs to address future challenges and how to increase rate of commercialization of novel reaction and separation technologies

American Institute of Chemical Engineers
3 Park Avenue, New York, NY
10016-5991

ISBN: 978-1-60423-471-8